RESEARCH AND PRACTICES IN WATER QUALITY

Edited by **Teang Shui Lee**

Research and Practices in Water Quality

http://dx.doi.org/10.5772/58512
Edited by Teang Shui Lee

Contributors

Toyin Arowolo, Nataša Gros, Hlanganani Tutu, Bronwyn Camden-Smith, Raymond Johnson, Peter Camden-Smith, Ty Yeh, Marco Andre Silva Costa, Magda Monteiro, Inna Rudenko, Sanaatbek Salaev, Sanjar Davletov, Ruzumboy Eshchanov, Wanshun Zhang, Innocent Rangeti, Bloodless Dzwairo, Fredrick Alfred Ochieng' Otieno, Graham. J Barratt, Petr Fučík, Teang Shui Lee, Olanrewaju Olusoji Olujimi

CBS, Edition **2017**

Published by InTech

Janeza Trdine 9, 51000 Rijeka, Croatia

Publishing Process Manager Iva Simcic

Technical Editor InTech DTP Team

Cover Designer InTech Design Team

Additional hard copies can be obtained from orders@intechopen.com

Research and Practices in Water Quality, Edited by Teang Shui Lee
p. cm.
ISBN 978-953-51-2163-3

INTECH

open science | open minds

Contents

Preface

Water quality is an important aspect of water resources, be it for portal water or for water prior to disposal into the various disposing channels and ultimately either into the lands or to the rivers and then open seas. As responsible scientists and engineers it is incumbent on us to ensure the quality of the environment and the water bodies are in their prime best at all times. The book "Research and Practices in Water Quality" contains twelve chapters, covering research and practical works about various aspects of modelling, guidelines, statistical research, as well as methods and practices in the field of water quality. These works should be of interest to those engaged in water quality, either in research on water quality or rivers and lakes, drawing guidelines for non-portable water quality, determining best statistical approaches to enhancing the filtering improvement for better water quality. In an era of droughts and floods undoubtedly the quality of water becomes significantly important no matter where the source of water is, if we are to keep at bay all the related diseases.

Prof. Teang Shui Lee
Professor of Water Resources Engineering,
Department of Agricultural and Biological Engineering,
Faculty of Engineering, Universiti Putra Malaysia
Malaysia

Section 1

Modelling, Guidelines and Statistical Research in Water Quality

Chapter 1

Applying Numerical Models for Water Environments in Watersheds – Case Studies of Tai Lake, Middle and Lower Han River and East Lake in China

Wanshun Zhang

Additional information is available at the end of the chapter

http://dx.doi.org/10.5772/60119

1. Introduction

The overall status of the global water environment has entered into a new stage exhibiting serious ecological degradation and complex environmental pollution. China is confronting critical water environmental problems, such as increasing pollutant loads and aggravated surface and ground water contamination. Based on the mechanisms of water environment evolution and contaminant transport, as well as theories of the watershed non-point source pollution model, the hydrodynamic model and water-quality model, this study built a comprehensive social-economic-hydrology-water environment model system for investigating water environments in watersheds. This model system can simulate the water cycle of watersheds and water environment quality changing processes, and analyse the utilization of water resources, pollutant discharges and the relationship between quantitative couplings and response connections, as well as reveal the formation mechanism of watershed pollutants and the laws of water environment revolution. This model system has successfully supported the comprehensive watersheds managements in Tai Lake, the middle and lower Han River and East Lake in China.

1.1. Research background

The load of pollutants entering the water environment has rapidly increased in China since the implementation of open and reform policies in China, while minor progress has been achieved in the mitigation of pollution. Among China's seven major river systems, the Songhua River and Huai River are slightly polluted, the Yellow River and the Liao River are moderately polluted and the Hai River is highly polluted. Eutrophication of lakes and reservoirs has become serious and is high in nitrogen and phosphorus. Drinking water may have been polluted with conventional contaminants and new types of toxic pollution, which threatens to health of urban and rural people.

The watersheds water environment model can be used in the simulation and evaluation of water environments, the forecasting and prediction of water quality, and can supplement the establishment of the standardization of pollutant discharges and water quality management. It is an important tool for the water planning, management and scientific research of watersheds. An established watersheds water environment model can describe the pollutant migration and transformation rules over time and in terms of space scale. We can make reasonable predictions about the development of the water environment based on the study of variables and fixed scientific parameters. Therefore, the study of watersheds water environment system simulation technology and especially the integration of other factors such as physical environment and ecology into the model system is one of the core technologies in watersheds water planning and environmental management.

Lake Tai is located in the Yangtze River Delta and is one of the five largest fresh water lakes in China. In China's lake district, many sub-lakes form extensive water networks. The Tai Lake basin is China's fastest changing and developing region. In recent years, due to the continuous degradation of water quality, eutrophication in Lake Tai has worsened and has had negative impacts on regional socio-economic development.

The total length of the Han River is 1532 kilometres, making it the largest tributary of the Yangtze River. Danjiangkou Reservoir, located in the middle and upper reaches of the Han River, acts as the water source of the South-To-North Water Transfer Project, which channels water from the Danjiangkou Reservoir through Henan and Hebei Province to the serious water shortage district in Beijing-Tianjin, China. It provides water to more than 100 million people along its main canal. The water transfer project reduced the reservoir's water level, substituting water resources supply and the demand balance of the middle and upper reaches of Han watersheds.

East Lake, located in Wuhan, is the second largest downtown lake in China; however, the lake has become atrophied. Due to a rapidly increasing urban population and intensification of the industrial and agricultural sectors, the lake has received excessive amounts of nitrogen and phosphorus, sourced from surrounding point- and non-point sources. The local authorities have made good progress on point-source pollution control. In order to achieve the overall ecological health of East Lake, the next step is to strengthen the control of non-point source pollution and establish a water management network for East Lake.

In summary, this study presents an adaptable watershed aquatic environment model that can be applied to Lake Tai, middle and lower of Han River, and East Lake. The study will establish: (1) an complex model of a river-lake network to achieve dynamic modelling of the Lake Tai water environment; (2) establish a comprehensive water quality model for a systematic water control project to quantitatively estimate the impact of water transfer; (3) a project for controlling the water quality of middle and downstream Han River; (4) build a non-point, dynamically-sourced 3D mathematically-coupled water quality model according to the characteristics of the received water and boundaries across watersheds, as well as the geography of East Lake. This approach will assist in answering how water contamination developed in each watershed, as well as the evolution patterns of the aquatic environment. The findings will help to improve control of the pollution of watersheds in order to maintain the sustainability of their environmental, social and economic development.

1.2. Research progresses

1.2.1. Non-point source model

The source of non-point pollution primarily derives from the application of fertilizers, pesticides, effluent irrigation and runoff from urban surfaces. Due to the complexity of these sources, the indetermination of the mechanism, it can be challenging to quantify the formation and load of non-point source pollution. In this instance, establishing models that stimulate watershed environments from time and space perspectives can be the most effective and direct measures. Usery et al. [1] coupled GIS and the distributed non-point source model to stimulate and evaluate watershed contaminants. Bryan [2] suggested a multi-standard evaluation measure in the study of non-point agricultural source pollution in Western Australia. Chowdary et al. [3] applied remote sensing and GIS to stimulate non-point agricultural pollutants and sediments on 2700 ha of land in Jharkhand State. Gikas et al. [4] utilized a SWAT model suitable for non-point source pollution in the Mediterranean region. Wang et al. [5] adapted LEAM, which was designated for modelling land use changes, to non-point source pollution and water quality modelling for the determination of the long-term effect of urbanization on water quality.

Recently, the intensification of non-point source pollution has drawn the increasing attention of the science community. Zhang [6-7] built a model of distributed urban-precipitation-runoff source pollution according to the properties of urban-sourced pollution and its transformation processes for the Project of Urban Water Environment Remediation Wuhan, which analysed pollution patterns resulting from different precipitations and land uses. Qiao et al. [8] coupled river network development DEM grid vector data to supply an alternative means for analysing basic terrain data in a non-point source model. Liu et al. [9] combined RS and GIS technology to construct a watershed non-point source pollution model to calculate pollution loads of various types of pollution sources. Tang et al. [10] applied a SWAT watershed non-point source pollution model to evaluate the effect of measures on water quality improvement in Wenyu River.

1.2.2. Water dynamics and quality model

Contaminants produced from lands are discharged as a single point source into rivers and lakes. The quantification of the migration and transformation of pollutants requires a dynamic water quality model. In light of the different properties of watersheds, such models study river network water dynamics and quality, including that of lakes and reservoirs.

1.2.2.1. *Dynamic river network water and quality model*

Many dynamic river network water and quality models have already undergone real-world applications. [11] Mikell [12] scattered the water level of watercourses and the flow of water into calculation points. Arega [13] established a simple two-dimensional waterflow-salinity model and applied obvious TVD limited volumetric calculus to solve the calculation of waterflow-salinity. Peng et al. [14] constructed the model for establishing the Han River's water quality ecology numerically, which took into account the effect of the top support action of the water level of the Yangtze River. They then numerically stimulated the migration and transformation of water dynamics, total phosphorus, dissolved oxygen and phytoplankton of the Han River.

1.2.2.2. *Lakes and reservoirs: water dynamics and quality model*

The common lake and reservoir water dynamics and quality models are grouped into zero-, one-, two- and three-dimensional modelling and according to specific resolutions. Vejzak [15] tested the effect of eutrophication on phytoplankton dynamic, built an eutrophication model and predicted the effect of different nutritional changes. Asaeda [16] stimulated the effect of large aquatic vegetation degradation on nutrient budget in shallow lakes. Kurup [17] applied the finite difference scheme, where TISAT and CE-QUAL-W2 were coupled and tested in a swan lake estuary. Angelini [18] utilized the ELLOBO model to illustrate three status varieties on reservoirs. AyseMuhammetoglu [19] suggested a three-dimensional model on the effect of advantageous large aquatic vegetation on the quality of water of shallow lakes. Wang et al. [20] applied sourced convection diffusion equation and the water ecology dynamic model, based on the two-dimensional watershed water flow-quality model, and as a result built the shallow water ecological restoration model, which they applied to the study of reservoir restoration in Shenzhen.

2. Non-point source model

2.1. Runoff-yield model

The sub-basin was considered as a nonlinear reservoir. We combined the Manning Continuity equation and established the nonlinear hysteresis of surface runoff storage. The Manning Continuity equation can be presented as:

$$\frac{dV}{dt} = A_{s1} \cdot \frac{dh}{dt} = A_{s1}P - Q_{W1} \tag{1}$$

where $As1$ is the basin area (m^2), V is sub-basin water rate (m^3), T is time (s), H is the water depth (m), $QW1$ is the sub-basin for the flow of traffic and P is net rain (m/s).

http://dx.doi.org/10.5772/60119

$$Q_{W1} = Q_{pl}^{'} + Q_{gl}^{'} \tag{2}$$

where $Q_{p1}{}'$ is the surface of the slope flow (m^3/s) and $Q_{g1}{}'$ is the flow into the ditch (m^3/s).

The flow equation for the sub-valley can be presented as:

$$Q_{W1} = W \cdot \frac{1}{n}\left(\text{h-h}_{\text{p}}\right)^{5/3} S^{1/2} \tag{3}$$

where W is the width of the sub-basin type (m), S is sub-basin slope, N is the Manning roughness surface and H_P is the hollow lag of storage water depth (mm).

The Von Hoyningen-Hune formula was applied for formulating canopy interception of rainfall:

$$P_i = aLAI\left[1 - \frac{1}{1 + \frac{\text{bp}_{gross}}{aLAI}}\right] \tag{4}$$

where Pi is the quantity of rainfall interception (mm), a is the empirical coefficient (cmd-1), pgross is rainfall (mm), LAI is the leaf area index (m^2m^{-2}) and b is soil coverage, b=$LAI/3$. With an increase in rainfall, the rainfall interception amount will gradually approached to the saturation value $aLAI$.

Fill the depression, in the form of incremental is:

$$\Delta V = \Delta P_e e^{-P_e/S_d} \tag{5}$$

where *Pe* is net rainfall (mm) and *Sd* is maximum filling depression (mm).

For infiltration, the Green-Ampt equation can be applied when initial rainfall is stronger than the infiltration amount, as well as when the initial rainfall is less than the infiltration amount. The surface infiltration capacity can be defined as:

when $F < F_s$:

$$f = i$$

$$F_s = \frac{S \cdot M}{i/K_s\text{-}1}$$

$$(i > K_s) \tag{6}$$

$$F_s = 0(i \le K_s),$$

when $F \ge F_s$:

$$f = f_p$$

$$f_p = K_s\left(1 + \frac{S \cdot M}{F}\right) \tag{7}$$

where f is infiltration rate (mm/s), F_p is stable infiltration rate (mm/s), M is the initial saturation (mm/mm), I is the rainfall intensity (mm/s), F is the cumulative infiltration amount (mm), K_s is soil saturated hydraulic conductivity (mm/s), S is the wetting front of capillary suction (mm) and F_s is saturated with the cumulative infiltration amount (mm).

2.2. The pollution-generation model

1. **Pollutant accumulation**

The cumulative equation can be represented as:

$$\frac{dL_{si}}{dt} = k_i\text{-}k_{2i}L_{si} \tag{8}$$

where i is the first type of surface coverage, K_i is dust sedimentation rate (g/m^2 day), L_{si} is the surface dust (g/m^2), K_{2i} is the consumption rate of dust fall (d^{-1}) and T is time (d).

The accumulated amount of dust, where *Lsi* is proportional to the accumulated amount of pollutants *Lij* can be presented as:

$$L_{si} = f_{ij} \cdot L_{ij} \tag{9}$$

where *fij* is the *i* kind of surface coverage on the characteristics of the j proportion coefficient of pollutants (mg/g).

2. **Pollutant washing**

The pollutant washing index equation can be represented as:

$$\Delta L_{ij} = L_{ij}\left[1 - \exp\left(\text{-}k_{3i}R\right)\right] \tag{10}$$

where ΔLij is the j type of pollutants (g/m^2) devided from the i cover characteristics of catchment units wash, Lij is the amount of pollutants accumulated at the start of rain in the j catchment units by covering i characteristics (g/m^2), $k3i$ is the scouring coefficient of the water unit by covering i characteristics set of (mm^{-1}) and R is rainfall (mm).

3. River network water dynamics and water quality model

We applied the Saint Venant equations to describe the process of river water dynamics; the fundamental equations can be presented as follows:

Continuity equation is,

$$\frac{\partial A_j}{\partial t}+\frac{\partial Q_j}{\partial x}=q_j \tag{11}$$

The momentum equation is,

$$\frac{\partial Q_j}{\partial t}+\frac{\partial u_j Q_j}{\partial x}+gA_j\frac{\partial z_j}{\partial x}+\frac{gn^2|u_j|Q_j}{R^{4/3}}=0 \tag{12}$$

where uj is the j river channel section's average flow velocity, Qj is the flow of river channel j, A is the crossing water area of river j, T is time, Qj is the river flow Q of lateral flow bus j, G is acceleration of gravity, zj is the j river channel water level, N is the river roughness and R is hydraulic radius.

The pollutants diffusion equation can be presented as:

$$\frac{\partial(h_j c_i)}{\partial t}+\frac{\partial(u_j h_j c_i)}{\partial x}=\frac{\partial^2(E h_j c_i)}{\partial x^2}-h_j k_d c_i+h_j S_m \tag{13}$$

where Ci is pollutant concentration, Hj is j river channel water level, uj is j river channel, E is the diffusion coefficient, Sm is pollutant source items, Kd is the pollutant degradation coefficient, $Kd=K0\theta^{T-20}$, T is the water temperature, θ is coefficient of 1 ~ 1.08 and $K0$ is the degradation coefficient of the normal temperature.

The river network node equation can be presented as:

$$z_{j_1}=z_{j_2}=\cdots=z_{j_n} \tag{14}$$

$$A_j \frac{\Delta z_{ji}}{\Delta t} = \sum Q_{ji} \tag{15}$$

$$c_s = \frac{\sum_{i_{in}} c_{i_{in}} Q_{i_{in}}}{\sum_{i_{out}} Q_{i_{out}}} \tag{16}$$

where *zj* is for the node j water level, *Aj* is the node j cross-sectional area, *cs* is the water concentration of node *j*, *ciin* is the water concentration of inflow and *Qiin* is the inflow flow.

4. Lake water dynamics and water quality model

We described the natural water body movement control equation using the continuity equation and momentum equation. Under the rectangular coordinate system, this can be expressed as follows:

Continuous equation is:

$$\frac{\partial h}{\partial t} + \frac{\partial hu}{\partial x} + \frac{\partial hv}{\partial y} = 0 \tag{17}$$

X direction momentum equation is:

$$\begin{aligned} &\frac{\partial hu}{\partial t} + \frac{\partial huu}{\partial x} + \frac{\partial hvu}{\partial y} + gh\frac{\partial z}{\partial x} + \frac{gn^2 h\sqrt{u^2+v^2}u}{h^{4/3}} = \\ &= hfv + \frac{\rho_a f_w \left(w_x^2 + w_y^2\right) w_x}{\rho_w} + \frac{\partial}{\partial x}(h\gamma_t \frac{\partial u}{\partial x}) + \frac{\partial}{\partial y}(h\gamma_t \frac{\partial u}{\partial y}) \end{aligned} \tag{18}$$

Y direction momentum equation is:

$$\begin{aligned} &\frac{\partial hv}{\partial t} + \frac{\partial huv}{\partial x} + \frac{\partial hvv}{\partial y} + gh\frac{\partial z}{\partial y} + \frac{gn^2 h\sqrt{u^2+v^2}v}{h^{4/3}} = \\ &= -hfu + \frac{\rho_a f_w \left(w_x^2 + w_y^2\right) w_y}{\rho_w} + \frac{\partial}{\partial x}(h\gamma_t \frac{\partial v}{\partial x}) + \frac{\partial}{\partial y}(h\gamma_t \frac{\partial v}{\partial y}) \end{aligned} \tag{19}$$

where u and v are the vertical average velocities on x and y direction, Z is the surface elevation, H is the depth of the water, F is Coriolis force coefficient $F = 2\Omega\sin\theta$, to which Ω is suitable for the earth's rotation angular frequency, θ is local latitude, γi is turbulence viscosity coefficient, ρa, ρw is the density of air and water, fw is wind stress coefficient and *Wx* and *Wy* are the wind speeds on x and y direction.

The pollutants diffusion equation can be presented as:

$$\frac{\partial hC_i}{\partial t}+\frac{\partial huC_i}{\partial x}+\frac{\partial hvC_i}{\partial y}=\frac{\partial}{\partial x}(hK_x\frac{\partial C_i}{\partial x})+\frac{\partial}{\partial y}(hK_y\frac{\partial C_i}{\partial y})-hS_i \tag{20}$$

where *Ci* is the concentration of contaminant i and *Kx* and *Ky* are the diffusion coefficients on the x and y directions. The pollutants diffusion equation fully considered that water convection and diffusion can potentially degrade pollutants.

5. One-dimensional and two-dimensional model coupling method

On the interface section of the one-dimensional and two-dimensional models, we used the one-dimensional model to simulate the river water level and change of flow rate, and took the data as an implicit variable of the two-dimensional lake model. At the same time, water level, flow and concentration showed equivalent conditions for both models, so that the coupling of one-dimensional and two-dimensional models was achieved. Transitional elements were the connecting units between the one-dimensional model and two-dimensional models. The transitional unit grid layout is shown in Figure 1.

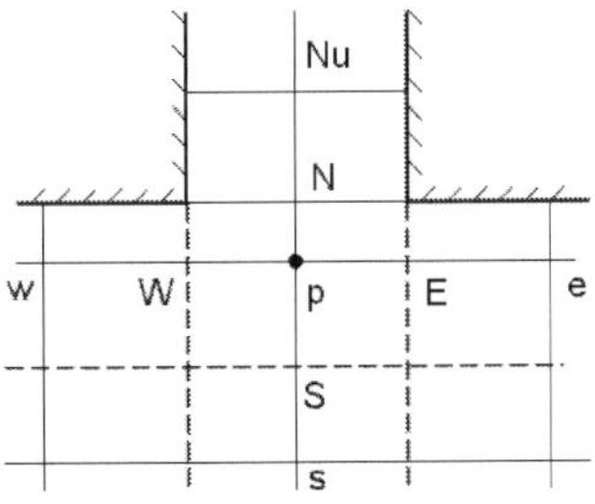

Figure 1. One-dimensional and two-dimensional models of connection.

Water connection condition:

$$Z_1 = Z_2 \tag{21}$$

Traffic connection conditions:

$$Q_1 = \int U_\varepsilon h_\varepsilon d\varepsilon \tag{22}$$

Water connection conditions:

$$C_1 = C_2 \tag{23}$$

where Z1, Z2 is the water level on the connection section of one- and two-dimensional models, C1, C2 is water quality concentration on the connection section of one- and two-dimensional models, Q1 is flow on the connection section of one- and two-dimensional models, *Us* is velocity on the connection section of one- and two-dimensional models, *uj* is the coordinate on the connection section of one- and two-dimensional models and H is the water depth.

6. Numerical dispersion and solving

Adapting the hydrodynamic and water quality model equation to a unified form can be presented as:

$$(\varphi)_t + (u\varphi)_x + (v\varphi)_y = (\varepsilon_s \varphi_x)_x + (\varepsilon_s \varphi_y)_y + S_\varphi \tag{24}$$

We used non-orthogonal and non-staggered grids to adopt the processing of the wind convection format within the control body, to integral and discrete on the formula(24), and to obtain the discrete equation of convection diffusion as follows:

$$a_p \varphi_p = \sum_{nb} a_{nb} \varphi_{nb} + S_p^\varphi \tag{25}$$

where ap and anb are the relevant coefficients, respectively.

We used the SIMPLE orthogonal algorithm to establish the η free surface correction equation and velocity correction equation. The η correction equation can be presented as:

$$d_p \eta'_p = \sum_{nb} d_{nb} \eta'_{nb} + S_p^\eta \tag{26}$$

The velocity correction equation can be presented as:

$$u'_p = u^*_p + b^u_p (\eta'_p - \eta'_e); v'_p = v^*_p + b^v_p (\eta'_p - \eta'_n) \tag{27}$$

where b_p^u, b_p^v, d_p and d_{np} are the discrete coefficients and u_p^*, v_p^* and η'_p are guesses.

The surface and velocity equations both belong to diagonal algebraic equations, where the SIMPLE algorithm can be applied for finding a rapid solution.

7. Examples of application

7.1. The survey of experimental research and the typical environmental processes in Lake Tai

7.1.1. The Lake Tai basin overview

Lake Tai basin is located in the delta plain area of Yangtze River and its watershed area is 36 500 km^2 [21]. Lake Tai basin is China's fastest changing and development region and its serious water pollution is a top concern. In investigating the complexity of the Lake Tai watershed, this study only selected Ge Lake, which is in the west of Lake Tai.

7.1.1.1. Natural environment conditions

The study area is located in upstream Lake Tai and is a major pollutant source of Lake Tai (as shown in Figure 2).The local climate is subtropical and experiences moist marine monsoons. It has abundant rainfall and sunlight hours annually. [22]

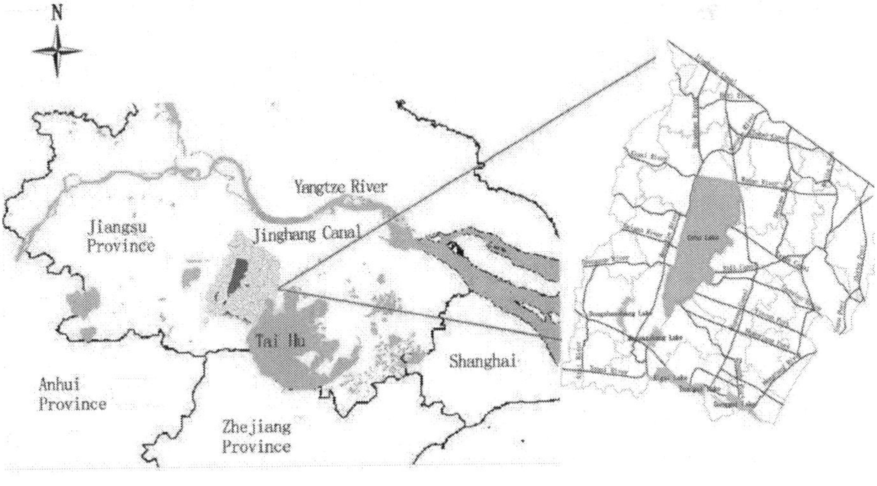

Figure 2. Location map of the Ge Lake river network of the Lake Tai Basin.

7.1.1.2. Social and economic situation

The local economy of the studied area is quite well-developed, with steady agronomy and rapidly increasing industrial economy. The local GDP per capita was 15,900 CNY in 2000 and 63 000 CNY in 2008, which was three times higher than the national average. [23]

7.1.2. Analysis and evaluation of water quality of Ge Lake

7.1.2.1. Water quality at Ge Lake

The results of water quality evaluation are shown in Table 1, where TP, TN and COD exceed largely when compared with the normal standard. The properties of organic pollution are highlighted.

Period	NH_3-N	COD	TN	TP	Water quality category
Dry season	Grade II	Grade V	Worse than Grade V	Grade III	Worse than Grade V
Normal season	Grade IV	Grade IV	Worse than Grade V	Grade IV	Worse than Grade V
Wet season	Grade II	Grade V	Worse than Grade V	Grade IV	Worse than Grade V

Table 1. Water quality evaluation results for Ge Lake in 2010.

7.1.2.2. The analysis of varying water quality in Ge Lake

The water quality is worse than Grade V at Ge Lake, with atrophied nutrition and the major pollution indicators of COD and NH3-N, as well as TN and TP. The lake water quality has declined from Grade III in the early 1990s to the poorer Grade V class and the lake's pollution index has risen 10% per year on average. Presently, algal species have become dominant and the lake has changed from being a grass-type lake into an algae-type lake.

7.1.3. Ge Lake water quality model

Combined with the features of the study area, in view of the mixture of pollutants present, a one-dimensional and two-dimensional mathematical model, respectively, were chosen to describe the hydrodynamic processes of the river and the transformation processes of pollutant migration. The finite control volume method was applied to solve the control equation of the two-dimensional hydrodynamic- and water-quality model (see Section 2). To generalize the research area, typical Ge-Lake watershed hydrology and water quality data were applied, which assisted in calibrating and verifying the model.

7.1.3.1. Drainage generalization

The river network in the study area was divided into different calculation sections, including 110 river sections, 91 calculation nodes and 519 calculation sections. The river network generalization results are shown in Figure 3.

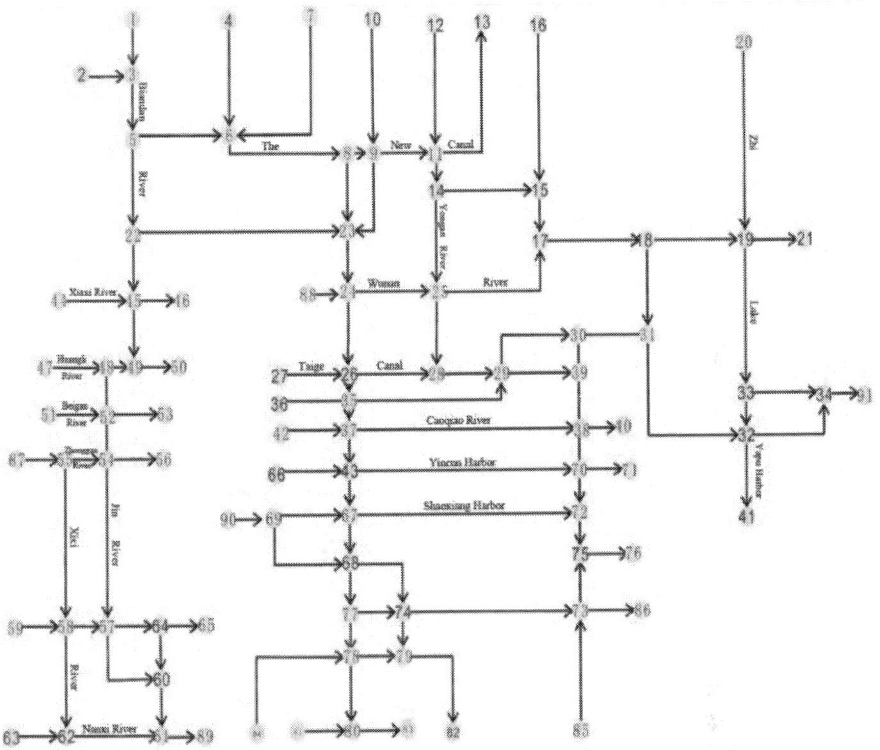

Figure 3. The typical river network profile in Ge-Lake.

7.1.3.2. Rate of model and validation

The distribution map of verification monitoring points in Ge Lake is shown in Figure 4.

Using daily water flow data from Xiaxi bridge station (Xiaxi River), Huangli station (Huangli River) and Dong'an bridge station (Beigang River) in 2007, we were able to calculate water outflow and input at every river sections. This was then applied to the verification of hydrodynamics at the upper-boundary conditions. Using daily water level data from Dukou station, Dapukou station and Yixing station at Lake Tai in 2007, we were able to calculate water output level at every river sections.This was then applied for the verification of hydrodynamics at lower-boundary conditions. We used pollution discharge load data in 2007 for the verification of the water quality model.

Figure 4. Distribution map of verification monitoring points in Ge Lake.

Correlation parameters

1. Roughness coefficient

The empirical formula and the result of the model parameter calibration determined the roughness coefficient n. River roughness was 0.02 ~ 0.025 and Ge-Lake had roughness of approximately 0.022.

2. Diffusion coefficient

The diffusion coefficient of the body of water in the study area (Ex) was obtained using an empirical formula according to Lagrange turbulence length and the intensity of turbulence length concept. The Ex values can be expressed as:

$$E_x = \alpha h u_* \tag{28}$$

$$u_* = \sqrt{ghI} \tag{29}$$

where α is the dimensionless coefficient, (0. 1-0.2); u_* is the friction velocity (m/s), I is the water surface slope and h is the water level (m).

3. **Degradation coefficient**

The degradation coefficient (K_0) of NH_3-N, COD, TN and TP are shown in Table 2.

Parameter	NH_3-N	COD	TN	TP
K_0 (d^{-1})	0.03-0.15	0.02-0.13	0.05-0.08	0.05

Table 2. Degradation coefficient (K_0) of NH_3-N, COD, TN and TP.

Model validation

1. **Hydrodynamic model validation**

Daily water flow and water level data (2007) from Huangnianqiao station (Taige Canal) and Caoqiao station (Caoqiao River) was used to verify the water dynamics of the model.

1. **flow verification**

The results of flow verification are shown in Figures 5 and 6. Overall, changes in flow calculation were consistent with changes in the measured data.

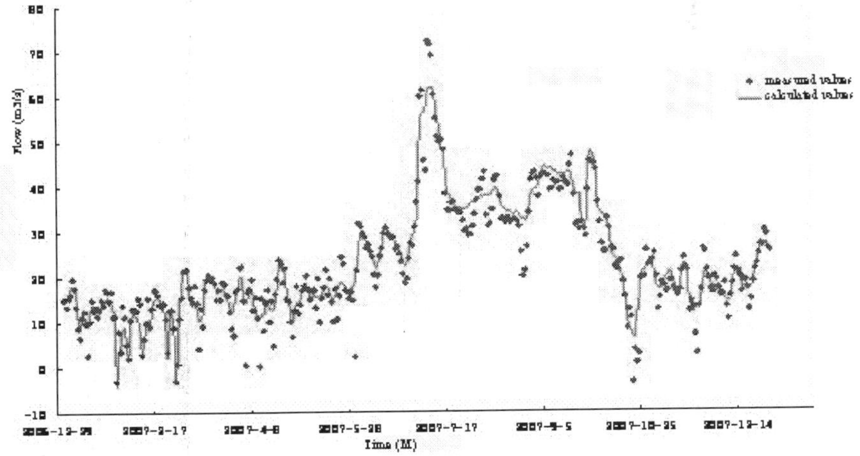

Figure 5. Flow verification of Huangnianqiao section.

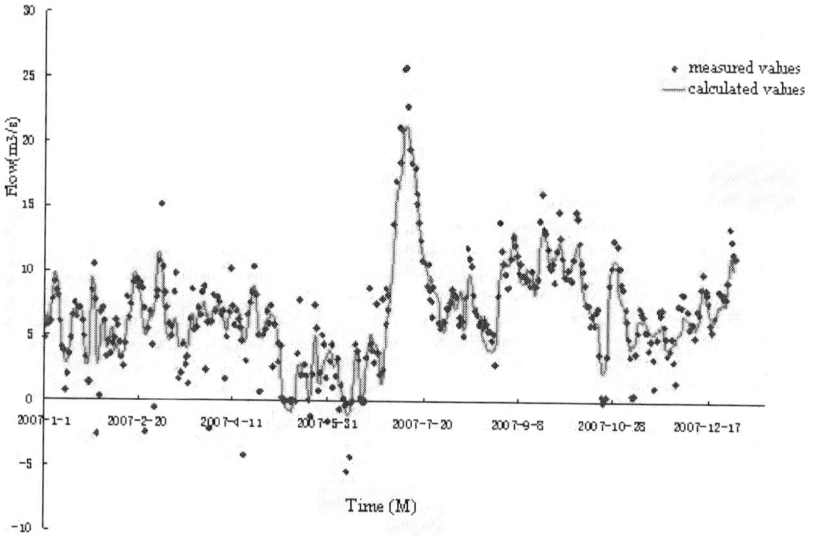

Figure 6. Flow verification of Caoqiao section.

2. Water level validation

The results of water level verification are shown in Figures 7 and 8. The precision of the model was good and it was able to meet the needs of the calculation.

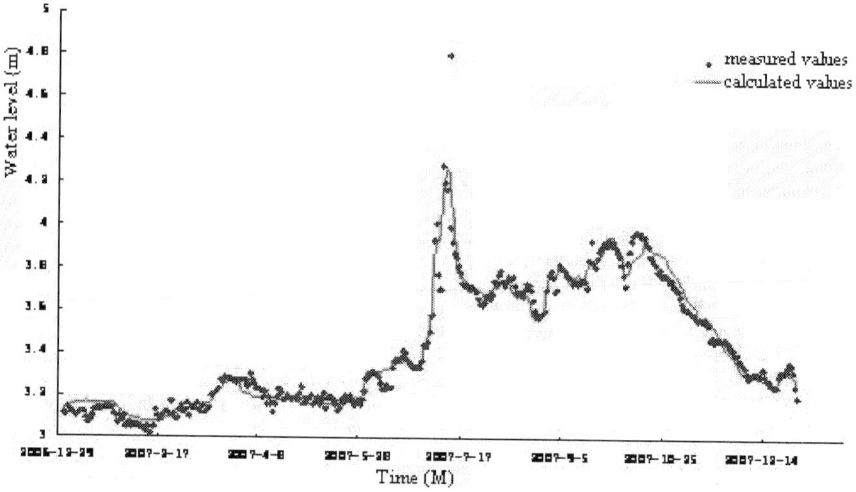

Figure 7. Water level verification of Huangnianhe section.

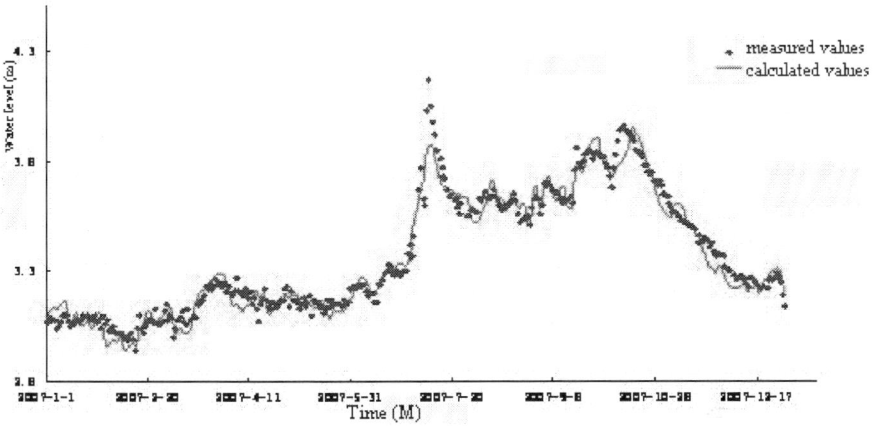

Figure 8. Water level verification of Caoqiao section.

The two-dimensional hydrodynamic model used daily water level data from Fangqian station (Ge Lake) for verification. It showed that the rule of calculated values and measured values had good consistency. Validation results are shown in Figure 9.

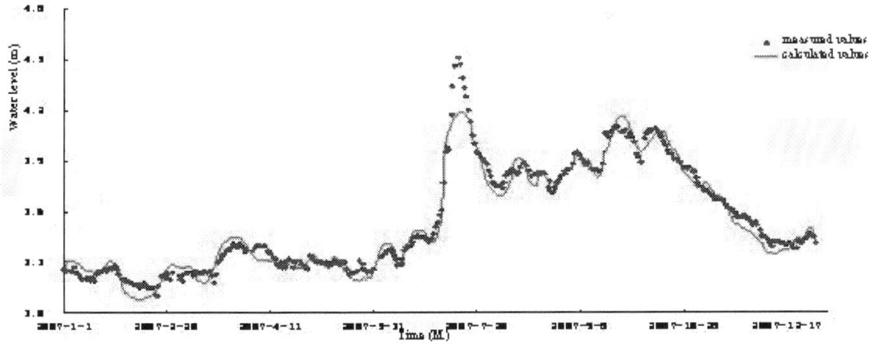

Figure 9. Water level verification of Ge-Lake.

3. Water quality model validation

Choosing three section stations within the scope of the study area, i.e., Huangnianqiao, Xuenianqiao and Cao Qiao, we used monthly-measured water quality data in 2007 to validate the river network model. The calculated results were compared with the measured values. Model simulated values and measured values were consistent, matching the requirements for the river's one-dimensional water quality simulation.

Sections	Relative Error			
	NH_3-N	COD	TN	TP
Huangnianqiao Section	23.98	18.92	23.49	21.56
Xuenianqiao Section	34.56	31.6	26.28	30.54
Caoqiao Section	20.5	16.68	22.65	23.56

Table 3. Average error of all section indexes (%).

We selected north Ge Lake as a regular water quality monitoring site. We validated the two-dimensional lake water quality model results by using measured data from January, April, July and October, 2007. As shown in Table 4, by comparing the simulated values and measured values, NH_3-N, COD, TN and TP in Ge Lake indicated that the relative errors were mostly less than 30% and the average error was roughly 24%.

Time	Relative Error			
	NH_3-N	COD	TN	TP
2007-1	18.92	20.11	14.66	17.39
2007-4	28.23	20.45	26.12	11.54
2007-7	29.03	21.55	13.24	24.48
2007-10	36.11	37.08	22	34.51

Table 4. Relative error of all indexes (%).

7.1.4. Conclusion

This study established a complex lake-river network coupled with a one- and two-dimensional water quality model under human interference. The research was based on socio-economic, demographical, climatic and hydrological data from 2007 at Ge Lake (Lake Tai region). Chemical oxygen demand, ammonium-N, total nitrogen and total phosphorus were selected as water quality indicators. The study analysed the water environment evolution rules as affected by human activities and interference. The results showed that ammonium-N and COD were mainly derived from domestic waste and industrial point sources; TN was primarily sourced from domestic waste and agricultural non-point sources; TP was largely derived from domestic and aquaculture pollution.

In view of the lakes in the region that were included in the river network, this study established a water environment holding capacity calculation model; the research findings are fairly adaptive to wider application in terms of water resource conservation and environmental management. The study also introduced a new angle for dynamic water-holding capacity modelling research. This approach can be integrated into pollutant control and environmental monitoring technologies in order to achieve the overall health of the water environment.

7.2. Modelling water quality and quantity according to the influence of cascade reservoirs and inter-basin water diversion projects on the middle and lower Han River [24]

7.2.1. Overview: the middle and lower Han River watersheds

At a length of 1531 km, the Han River is the largest tributary of the Yangtze River. The middle- and lower-Han rivers begin at the downstream end of the Danjiangkou Reservoir and flows for about 650 km before joining the Yangtze River at Wuhan City (see Figure 10). The total area of the drainage basin is about 63 800 km2. The river's reaches have been extensively dammed since the 1950s and its course has been highly regulated by a series of coupled reservoirs, among which Danjiangkou Reservoir is the largest; also contributing to this is the water source for the Middle Route Project (MRP) for the South-to-North Water Diversion (SNWD) Project, about 9.5 billion m^3 of water will be extracted from the Danjiangkou Reservoir and supplied to northern China annually, amounting to about 30% of the current annual downstream discharge. In order to mitigate the project's effects, the Yangtze-Hanjiang Water Diversion Project (YHWD) is supposed to transfer water from the upstream of the Yangtze River to the downstream of the Hanjiang River near Qianjiang City with a maximum capacity of 500 m^3/s.

The cascade reservoirs and the inter-basin water diversion projects, i.e. the MRP and the YHWD, lead to profound and complex effects in the middle and lower Hanjiang River

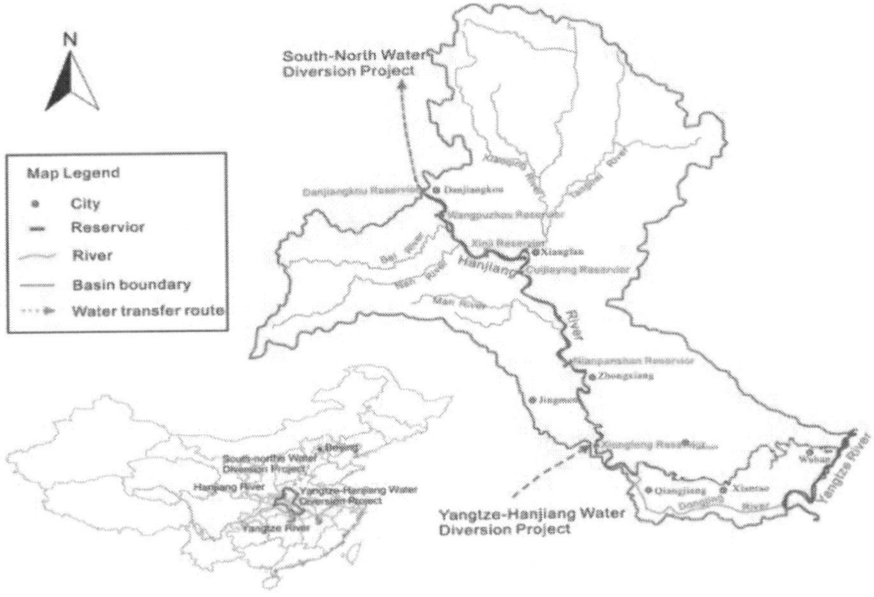

Figure 10. Location map for the middle and lower Han River Basin.

7.2.2. *Application*

7.2.2.1. *Model set-up*

A coupled one-dimensional hydrodynamic- and water-quality model was developed in a bid to assess water quality and quantity in the middle- and lower-Han rivers as a result of the impact of the operation of the cascade reservoirs and inter-basin water diversion projects. The systematic model includes numerical descriptions of the hydrodynamic and pollutant transport processes, the loads of different water users, stream-aquifer interactions, as well as the operations of cascade reservoirs and inter-basin water transfer projects (see Chapter 2).

7.2.2.2. *Division of the study basin*

The middle and lower Han rivers were divided into five reaches with 157 sections and the area outside the river was divided into 18 sub-hydrological units. A schematic illustration of the watershed is shown in Figure 11.

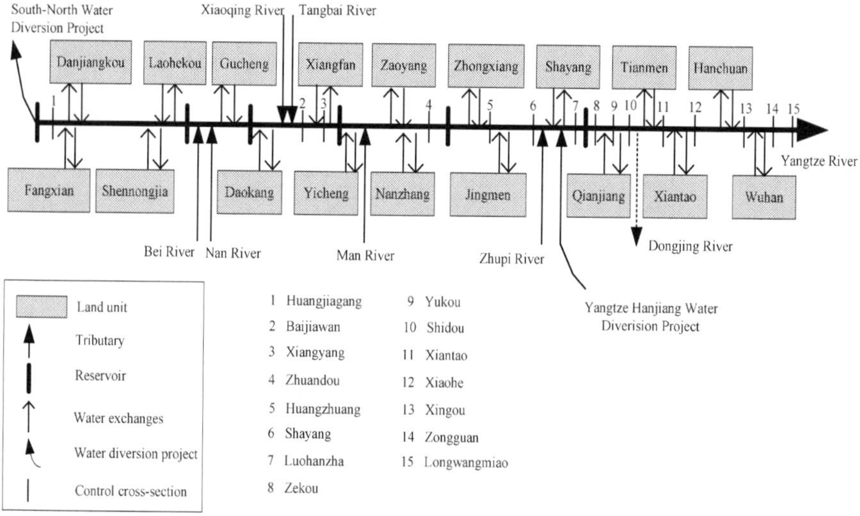

Figure 11. Schematic illustration of the basin.

7.2.2.3. *Model calibration and validation*

1. **Parameters**

Manning's roughness coefficient n1d took on the values of 0.03 to 0.055 according to a hydrological report from the Department of Water Resources of Hubei Province.

The value of the longitudinal dispersion coefficient ranged from 0.05 to 1 m^2/s, according to the national standard of the Technical Guide and Standard of Environmental Impact Assessment issued by China's Ministry of Environmental Protection. The degradation coefficients of pollutants are shown in Table 5.

Pollutant	COD_{Mn}	NH_3-N	TP
Degradation coefficient	0.25-0.3	0.33-0.35	0.06-0.09

Table 5. Degradation coefficients of different pollutants (d^{-1}).

2. Calibration and validation

The monitoring data of hydrodynamics and water quality from 2005 were used for calibrating the model. COD_{Mn}, NH_3-N and TP were chosen as water-quality indexes. The hydrodynamics and water quality were simulated along five monitoring sections using data from 2006 to 2007 to validate the model. The average relative deviations of the water levels between the simulated values and the measured values were within 1% of each other. The average relative deviations of the computed COD_{Mn}, NH_3-N and TP concentrations vs. the measured values were less than 20%. The model could therefore accurately simulate hydrodynamic conditions and water quality.

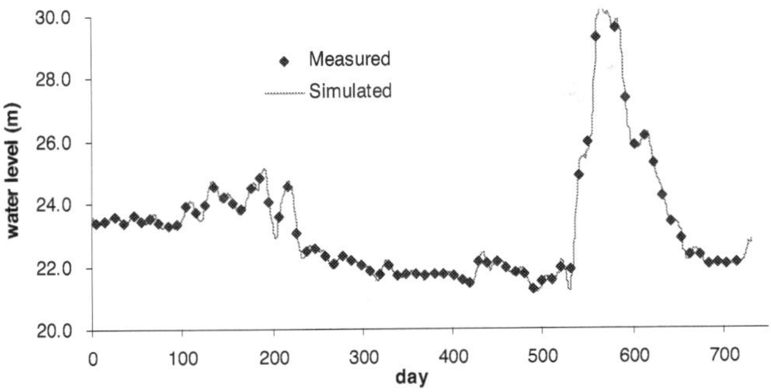

Figure 12. Hydrodynamic validation results of typical control cross-section 3.

7.2.3. Results

The long-term series of hydrological data from 1956 to 1998 were analysed to evaluate the long-standing trends in water quantity and quality of the middle and lower Han rivers. Two hydrological conditions were involved: the existing hydrological conditions prior to starting inter-basin water division projects and future hydrological-hydrodynamic conditions considering the cascade reservoirs and inter-basin water diversion projects (after initiating the projects).

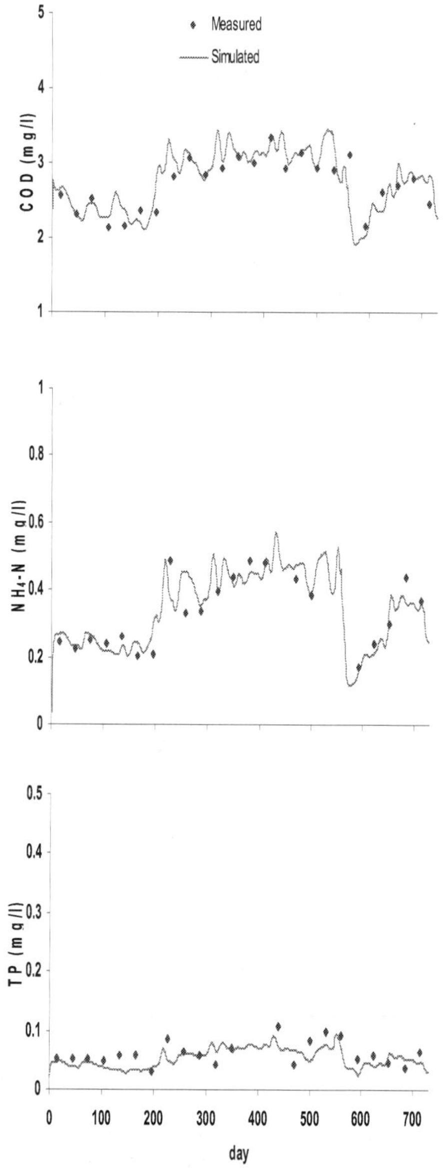

Figure 13. Water-quality validation results of typical control cross-section 3 (calculated and measured concentration of COD_{Mn}, NH_3-N and TP).

7.2.3.1. Analysis of water quantity

Based on the daily water demand/supply balance calculation, the discharge volume along the stream of the middle and lower Han rivers was analysed. Six sections in the middle and lower reaches were selected as control sections for six major cities along the river. Sections 1, 3, 5, 6, 8 and 13 respectively, represented the conditions of Danjiangkou, Xiangfan, Jingmen, Shayang, Qianjiang and Wuhan reaches.

Table 6 shows the average annual discharges of the six control sections prior to and following the projects. According to the calculated results, the problem of water deficiency will be serious for the middle reaches of the Han river in the future, as the operation of the MRP will significantly decrease stream discharge. As shown in Tables 7 and 8, operation of the cascade reservoirs and the MRP will significantly impact the hydrological/hydrodynamic conditions of the middle and lower Han rivers. The average water level will be decreased by 0.23 to 0.39 m in the river reaches downstream of Danjiangkou Reservoir, except for section 3 and section 6, which are controlled by the operation of Cuijiaying Reservoir and Xinglong Reservoir.

Control section	Before the projects (m^3/s)	After the projects (m^3/s)	Reduction
1	1161.6	811.3	-350.3
3	1292.7	912.7	-380.0
5	1518.0	1157.2	-360.8
6	1497.8	1139.3	-358.5
8	1341.5	1195.2	-146.3
13	1277.0	1140.0	-137.0

Table 6. Comparison of the average annual discharge of the control sections before/after the SNWD and YHWD projects.

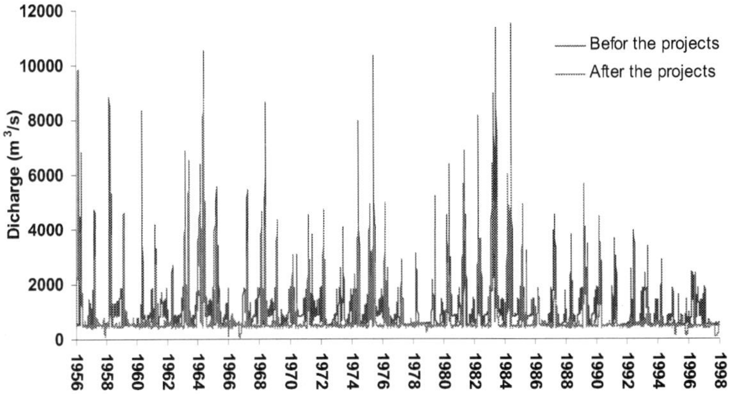

Figure 14. Calculated daily discharge of section 1 before/after the projects.

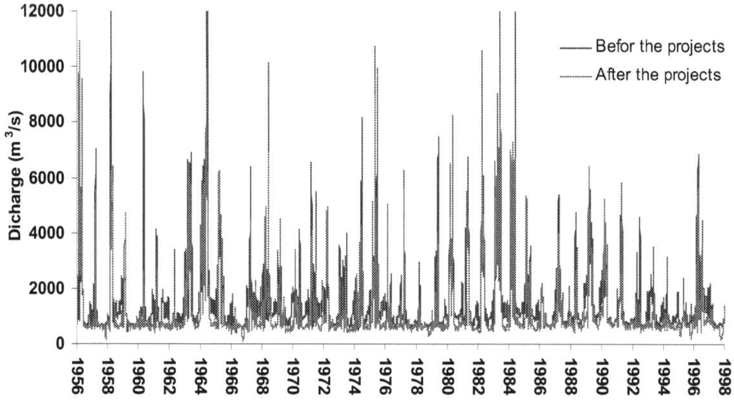

Figure 15. Calculated daily discharge of section 3 before/after the projects.

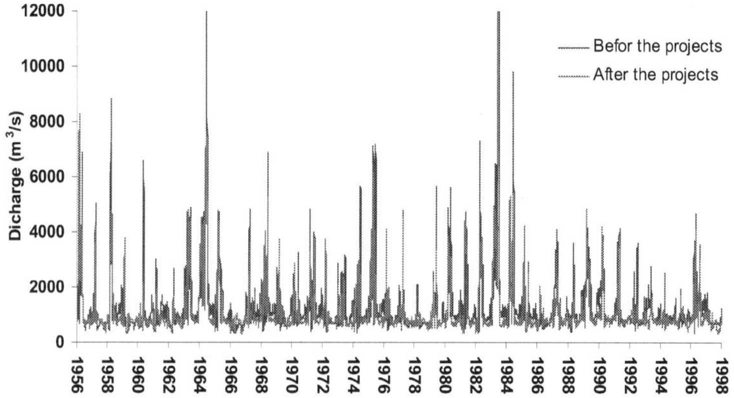

Figure 16. Calculated daily discharge of section 13 before/after the projects.

Control section	Before the projects (m)	After the projects (m)	Reduction
1	87.23	86.85	-0.38
3	58.67	62.78	4.11
5	40.04	39.65	-0.39
6	33.47	36.35	2.88
8	29.5	29.27	-0.23
11	24.29	24.03	-0.26

Table 7. Comparison of the average annual water level of the control sections before/after the SNWD and YHWD projects.

Control section	Before the projects (m/s)	After the projects (m/s)	Reduction
1	0.73	0.59	-0.14
3	0.9	0.31	-0.59
5	1.11	0.97	-0.14
6	0.69	0.32	-0.37
8	0.74	0.72	-0.02
11	0.67	0.64	-0.03

Table 8. Comparison of the average annual flow velocity of the control sections before/after the SNWD and YHWD projects.

7.2.3.2. Analysis of water quality

The results showed that the middle-lower reaches of the Han river will have similar trends of water-quality degradation. The average concentration of COD_{Mn} in the river was 2.61 mg/L, NH3-N 0.52 mg/L and TP 0.16 mg/L with existing hydrological-hydrodynamic conditions; with future operation of the cascade reservoirs and inter-basin water diversion projects, the average concentration of COD_{Mn} will increase by 55% to 4.04 mg/L, NH_3-N by 48% to 0.78 mg/L and TP by 46% to 0.24 mg/L.

Following the completion of the projects, the water quality will be degraded to Class IV. This result indicates that the operation of the YHWD project will be unable to mitigate water-quality deterioration in the lower Han River. The model simulations indicated that the dual effects of the projects, i.e., increasing the nutrient concentration and decreasing the flow velocity, could induce an environment favourable for algal growth. Thus, large-scale blooms likely occurred, even if only in the middle reaches of the river.

Sections	COD_{Mn}		NH_3-N		TP	
	Before the projects	After the projects	Before the projects	After the projects	Before the projects	After the projects
1	2.08	3.09	0.14	0.22	0.03	0.04
2	3.18	4.67	0.75	1.09	0.14	0.20
4	3.04	4.47	0.68	1.00	0.15	0.22
5	2.97	4.38	0.61	0.91	0.16	0.23
6	2.65	3.99	0.54	0.82	0.16	0.23
7	2.62	3.95	0.53	0.81	0.16	0.23
8	2.38	4.04	0.47	0.73	0.16	0.24
9	2.30	3.89	0.45	0.70	0.15	0.24
10	2.17	3.65	0.42	0.65	0.17	0.24
11	2.47	3.94	0.48	0.71	0.17	0.25
12	2.51	3.86	0.48	0.68	0.17	0.25
13	2.49	3.83	0.47	0.68	0.17	0.25
14	2.66	3.92	0.47	0.67	0.18	0.26
15	2.47	3.98	0.47	0.67	0.18	0.26

Table 9. Comparison of the simulated water quality of the main sections before/after the SNWD and YHWD projects.

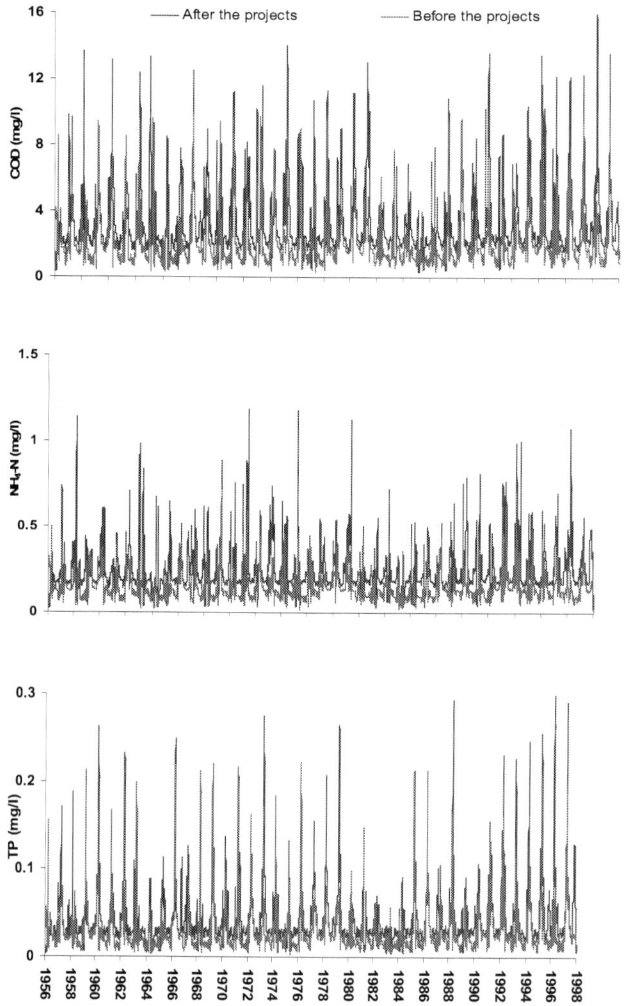

Figure 17. Calculated daily concentrations of COD, NH_3-N and TP of section 4 before/after the projects

7.2.4. Conclusions

A combined water-quantity-quality model was applied to simulate the current and future characteristics of water quantity and water quality in the middle and lower Han rivers. The simulation results showed that the implementation of the MRP and the operation of cascade reservoirs will exacerbate the problems of water pollution and water deficiency in the middle and lower reaches of the river. This model can assist water resource managers to make the right decisions about water improvement measures.

7.3. East Lake experimental research survey and typical environmental processes

7.3.1. The natural environment

East Lake is constituted by Guozheng Lake, Hou Lake, Miao Lake, Shuiguo Lake, Tangling Lake and others. Its geological structure is complex; the length of East Lake is 11.5 km and its average width 2.9 km. East Lake is connected to the Yangtze River via the Castle Peak port.

East Lake lies in Wuchang, which is the political, cultural and information centre of Hubei Province.

7.3.2. The non-point source water quality model

We built the non-point source model based on the digital elevation map (DEM). On the basis of hydraulics and the water quality model, we established the East Lake three-dimensional hydrodynamic and water quality model equation (see Chapter 2).This scenario was based on 2006 calibration results. After removing the point sources, we used the non-point source data to simulate the water quality status of East Lake from 2007 to 2010.

7.3.2.1. Division of the study basin

East Lake was divided into five reaches and the quantity of grid measured as 47 x 62; a schematic illustration of the watershed is shown in Figure 18.

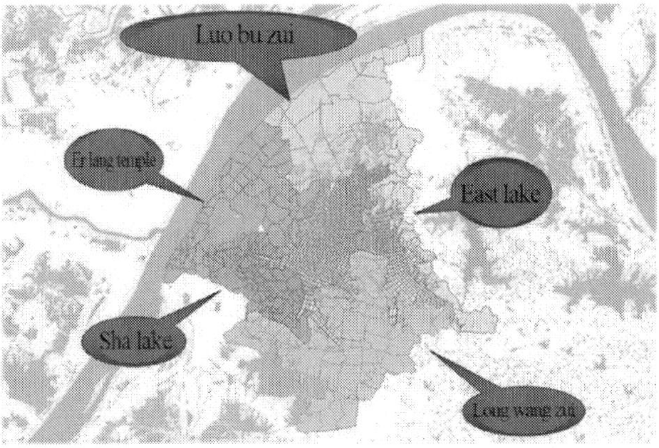

Figure 18. The division map for East Lake.

7.3.2.2. Correlation parameters

According to the measured data of East Lake from 2000 to 2005, we calibrated parameters, such as roughness coefficient, nitrogen degradation coefficient, organophosphate degradation coefficient in this study.

7.3.2.3. Computational conditions and pollution load input values

Using the measured results of water level and water quality from 2006 as initial conditions for control within the model, we calculated the changing process of flow, water level, water temperature, as well as the water quality of TP, TN, NH_4, DO.

Through a survey about the pollution circumstances of East Lake, we confirmed 22 outlets surrounding the main point source and 61 non-point source outlets. Their positions are shown in Figure 19.

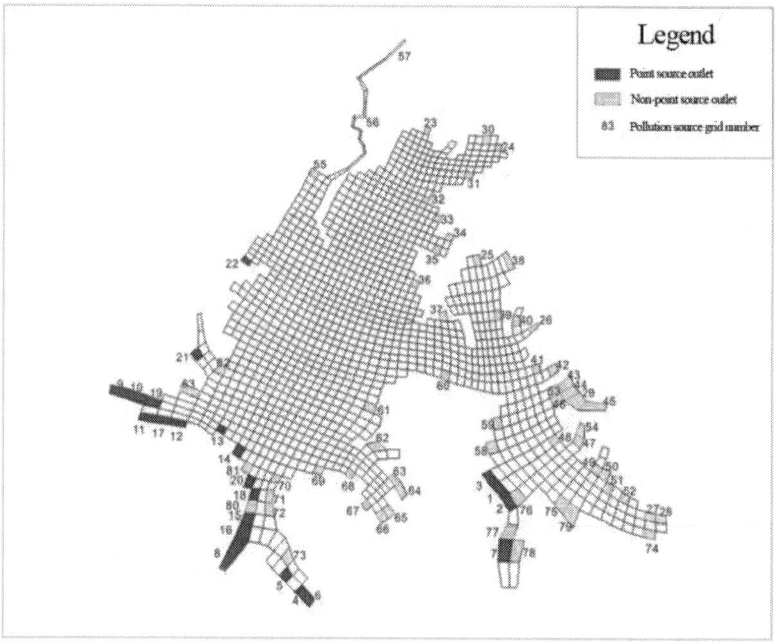

Figure 19. The source of pollutants map for East Lake.

7.3.2.4. Model validation

In this research, we used measured data of East Lake from 2006 to verify the model.

http://dx.doi.org/10.5772/60119

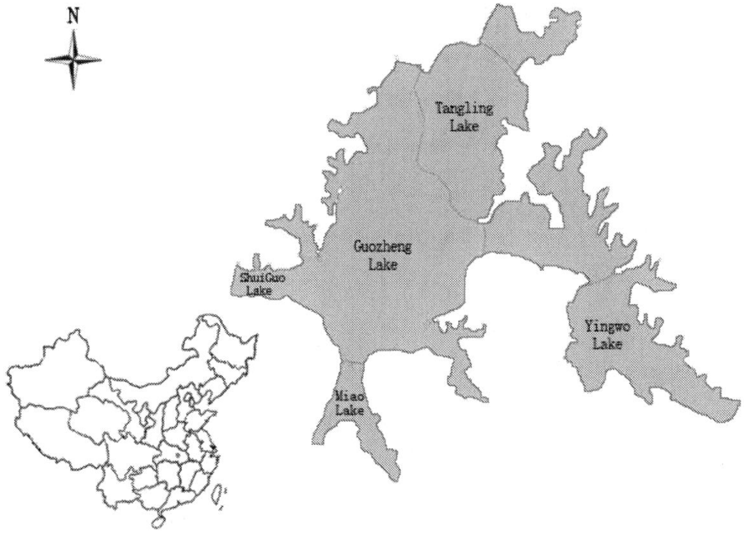

Figure 20. Location map of East Lake.

The model was verified as follows:

1. Results for Miao Lake verification are shown in Figures 21 to 24 below:

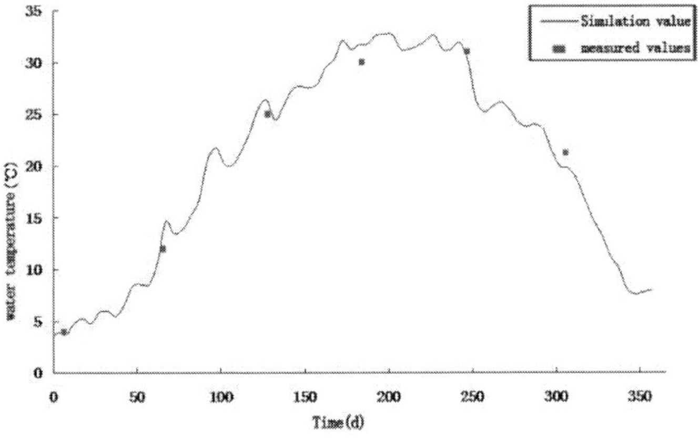

Figure 21. Verification chart for Miao Lake temperature.

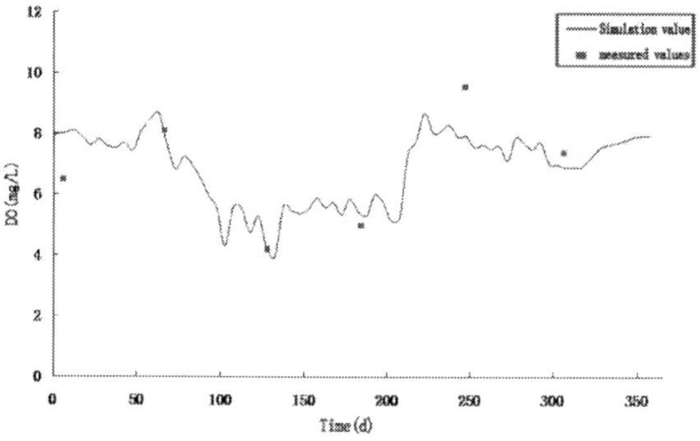

Figure 22. Verification chart for Miao Lake DO.

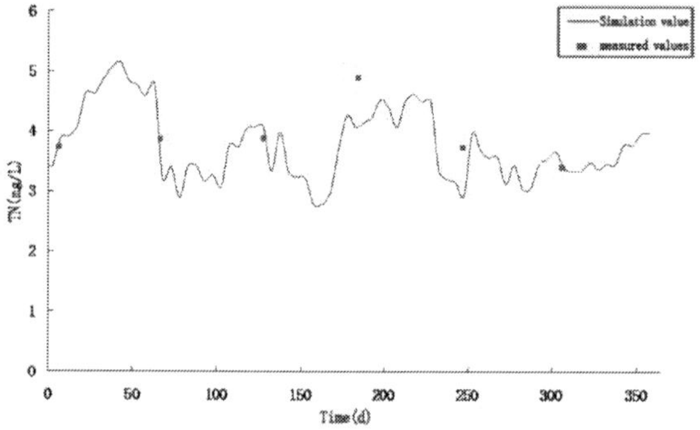

Figure 23. Verification chart for Miao Lake TN.

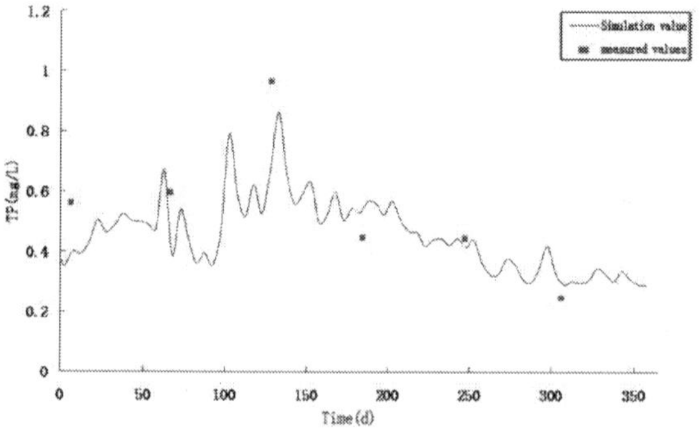

Figure 24. Verification chart for Miao Lake TP.

2. The verification results for Tangling Lake are shown in Figures 25 and 26 below:

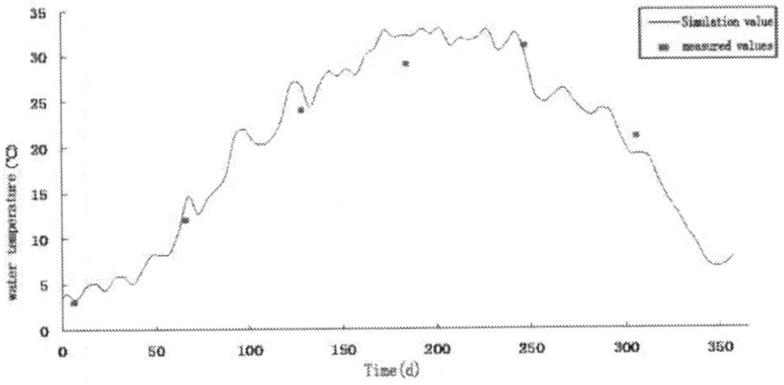

Figure 25. Verification chart for Tangling Lake temperature.

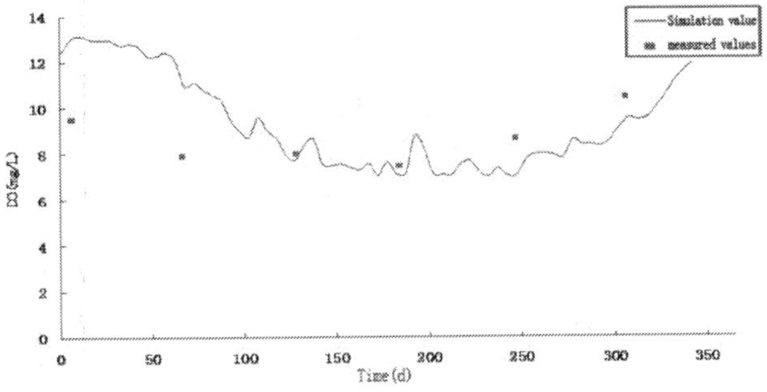

Figure 26. The verification chart for Tangling Lake DO.

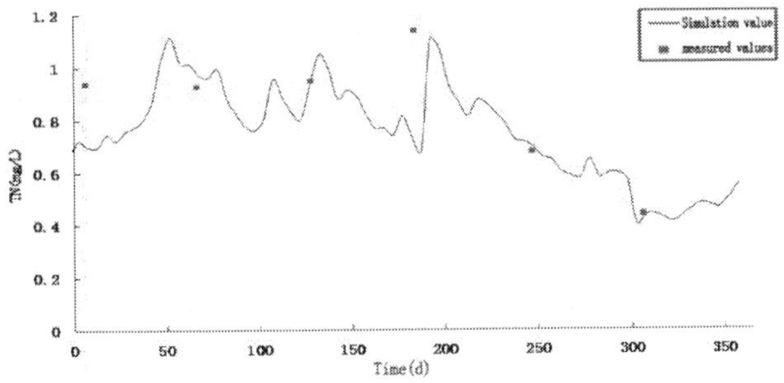

Figure 27. The verification chart for Tangling Lake TN.

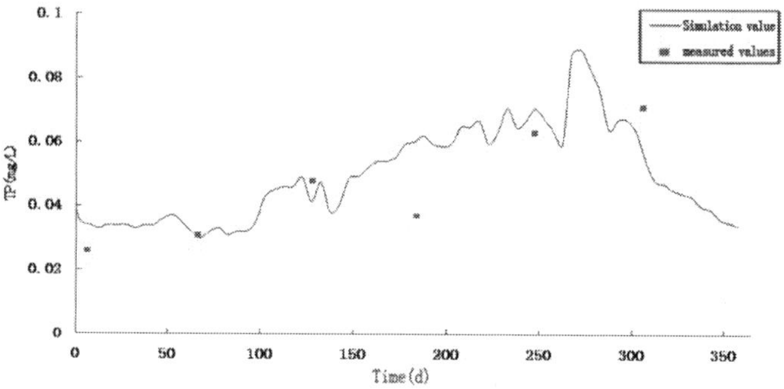

Figure 28. Verification chart for Tangling Lake TP.

The verification charts show that the maximum error of simulated calculation was less than 20% when compared to the measured concentration. The relative error in about 96% of samples was less than 10%. Results showed that the model was able to meet the requirements of simulation research.

7.3.2.5. Model application

The simulation results showed that the management of point sources will improve the water quality of East Lake, but that the effect will not be obvious in the short-term. The management of non-point source pollution should also be strengthened.

We conducted our research by including two study lakes. The results showed that the water quality of Miao Lake improved significantly, while Tang Ling Lake's did not.

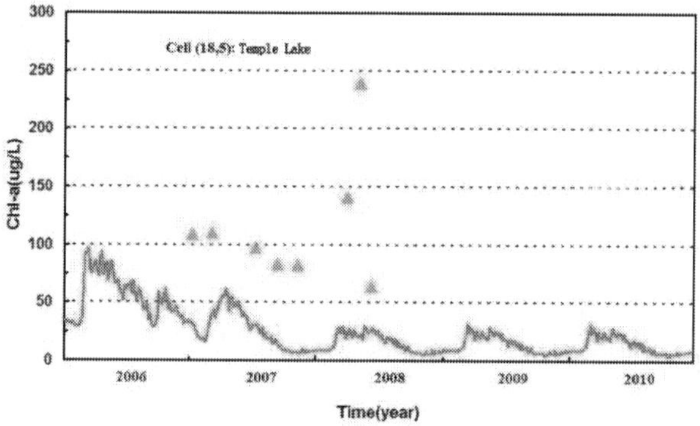

Figure 29. Water quality simulation of Miao Lake.

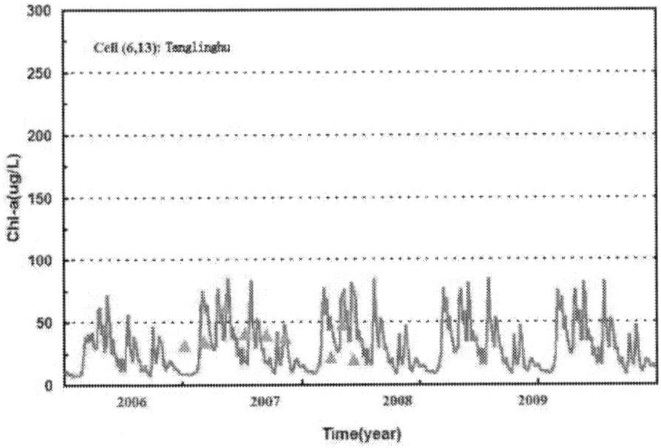

Figure 30. Water quality simulation of Tangling Lake.

7.3.3. Conclusion

The present research established a three-dimensional layered water hydrodynamic numerical model of East Lake. Using the measured data of hydrology and water quality from 2006, we completed the model parameters for calibration and validation. Then, we analysed the water quality and drew the following conclusion: the water quality of East Lake region must be improved. Factors such as non-point source and sediment need to be considered in the overall consideration.

8. Conclusion

The watersheds water environment system model established in this study has a solid theoretical basis and was successfully applied in the water environment response and simulation analysis of Tai Lake, the middle and lower Han rivers and East Lake.

1. We established a two-dimensional interfaced complex model of the river-lake network coupled with water quality, which was applied to the river network of Lake Tai using a series of conditions related to social economy, population, meteorological and hydrological factors measured in 2007. The research findings are fairly adaptive to wider application in terms of water resource conservation and environmental management.

2. We used the coupled water-quantity-quality model to simulate the current and future characteristics of water quantity and water quality for the middle and lower Han rivers. The model can assist administrators decision-making concerning water improvement measures.

3. Based on the hydraulic and water quality models, we created a three dimensional hydrodynamic and water quality model that can be successfully applied to the simulation of non-point source pollution, thereby providing a scientific tool for water environment management work.

Author details

Wanshun Zhang*

Address all correspondence to: wszhang@whu.edu.cn

School of Resources and Environmental Science, Wuhan University, Wuhan, China

References

[1] Usery EL, Finn MP, Scheidt DJ, Ruhl S, Beard T, Bearden M. Geospatial data resampling and resolution effects on watershed modeling: A case study using the agricultural non-point source pollution model. Journal of geographical systems 2004;6(3): 289-306.

[2] Bryan BA, Kandulu JM. Designing a policy mix and sequence for mitigating agricultural non-point source pollution in a water supply catchment. Water resources management 2011;25(3): 875-892.

[3] Chowdary VM, Yatindranath, Kar S, Adiga S. Modeling of non-point source pollution in a watershed using remote sensing and GIS. Journal of the Indian Society of Remote Sensing 2004;32(1): 59-73.

[4] Pisinaras V, Petalas C, Gikas GD. Gemitzi A, Tsihrintzis VA. Hydrological and water quality modeling in a medium-sized basin using the Soil and Water Assessment Tool (SWAT). Desalination 2010;250:274-286.

[5] Wang Y, Choi W, Deal BM. Long-term impacts of land-use change on non-point source pollutant loads for the St. Louis metropolitan area, USA. Environmental Management 2005;35(2): 194-205.

[6] Qiao F, Meng W, Zhang BH, Lei K, Zhang WS, Wang Y. Study of dividing slope of catchment based on DEM. Science of Surveying and Mapping 2011;3(2):26-28.

[7] Liu Z, Li WX, Zhang YM, Zhang LJ, Zhang HL, Li Y, et al. Estimation of Non-point Source Polltion Load in Taihu Lake Basin. Journal of Ecology and Rural Environment. 2010; 26(1):45-48.

[8] Zhang ML, Shen YM. Study on the Hydrodynamic and Water Quality Model in River. Doctoral Dissertation of Dalian University of Technology 2007.

[9] Arega F, Hayter E. Coupled consolidation and contaminant transport model for simulating migration of contaminants through the sediment and a cap. Applied Mathematical Modeling 2008;09:2413-2428.

[10] Vezjak M, Savsek T, Stuhler EA. System dynamics of eutrophication processes in lakes. European Journal of Operational Research 1998;109:442-451.

[11] Asaeda T, Trung VK, Manatunge J. Modeling the effects of macrophyte growth and decomposition on the nutrient budget in Shallow Lakes. Aquatic Botany 2000;68:217-237.

[12] Kurup RG, Hamilton DP, Phillips RL. Comparison of two 2-D, laterally averaged hydrodynamic model applications to the Swan River Estuary. Mathematics and Computers in Simulation 2000;51:627-638.

[13] Angelini R, Petrere M.A model for the plankton system of the Broa River. Ecological Modeling 2000;126:131-137.

[14] Muhammetoglu A, Soyupak S. 3-D water quality macrophyte interaction model for shallow lakes. Ecological Modeling 2000;133:161-180.

[15] Xiao C. Simulation and Prediction Research on Distributed Urban Non-point Source Pollution of Rainfall-Runoff. Doctoral Dissertation of Wuhan University 2005.

[16] Liu HX. Wuhan Research-point source pollution model. Doctoral Dissertation of Wuhan University 2009.

[17] Tang LH, Zhang SC, Lin WJ, Wang PJ. Scenario analysis of water pollution control in Wenyu watersheds in Beijing. Journal of Hydroelectric Engineering 2012;4:156-161.

[18] Zhou HC, Zhang GH, Wang GL. Multi-objective decision making aprroach based on entropy weights for reservoir flood control operation. Shui Li Xue Bao 2007;38(1): 100-106.

[19] Peng H, Guo SL. The Ecological water quality model and numerical simulation of the downstream of the Han River. Resources and Environment in the Yangtze Basin 2002;11(4):363-369.

[20] Peng H, Wang Y, Zhang WS, Chen WX, Feng TH. The numerical simulation of the shallow water ecological restoration. Yangtze River 200738(1):98-100.

[21] Zhu GW. Eutrophic status and causing factors for a large, shallow and subtropical Lake Taihu, China. Journal of Lake Sciences 2008;20(1):21-26.

[22] Li W. Does Lakes Environmental Kuznets Curve Exist: Case of Tai-lake. Ecological Economy 2014;1:363-370

[23] Zhao YX. Modeling water quality and quantity with the influence of cascade reservoirs and inter-basin water diversion projects in Middle and Lower Hanjiang River. Manuscript submitted for publication.

[24] Huang DF, Yang SZ, Liu ZQ, Mei ZY. Geological Studies of the Formation and Development of The Three Large Fresh-Water Lakes in The Lower Yangtze Valley. Oceanologia Et Limnologia Sinica 1965;11:396-426.

Chapter 2

Geochemical Modelling of Water Quality and Solutes Transport from Mining Environments

Bronwyn Camden-Smith, Raymond H. Johnson,
Peter Camden-Smith and Hlanganani Tutu

Additional information is available at the end of the chapter

http://dx.doi.org/10.5772/59234

1. Introduction

Water quality has become more and more important over the years owing to increased demand for water by the growing population. This has created some strain on the supply side as most water resources have become polluted, largely through anthropogenic activities such as agriculture, mining, manufacturing and domestic and industrial effluent among others. In South Africa, for instance, mining accounts for the largest part of water pollution due, largely, to abandoned mine sites that have, over the years, continued to release contaminants into surface and groundwater unabated.

The major challenge that abandoned mine sites, for example gold and coal mines, pose is that of acid mine drainage (AMD), a phenomenon that has been well documented [1-3]. There are four sub-reactions involved in AMD [1,4], the first reaction being the oxidation of sulphide to sulphate and ferrous ion:

$$2FeS_2 + 7/2O_2 + H_2O \rightarrow Fe^{2+} + 4SO_4^{2-} + 4H^+ \quad (1)$$

The resulting ferrous iron undergoes oxidation as it enters receiving surface water systems. This reaction is dependent on pH and oxygen and proceeds slowly under acidic conditions (pH< 3):

$$4Fe^{2+} + O_2 + 4H^+ \rightarrow 4Fe^{3+} + 2H_2O \quad (2)$$

This reaction is catalysed by the acidophilic *Thiobacillus ferrooxidans* bacteria. The slow rate of this reaction makes it the overall rate-determining step for the acid-generating process. The emanating acidity (H^+) can under neutralising reactions on contact with minerals such as calcite ($CaCO_3$), resulting in elevated pH and sustained concentrations of $Fe^{2+}$and SO_4^{2-}.

Ferric ion (Fe^{3+}) produced in reaction 2 precipitates as ferric hydroxide ($Fe(OH)_3$), releasing more acidity:

$$4Fe^{3+} + 12H_2O \rightarrow 4Fe\left(OH\right)_3 \downarrow + 12H^+ \quad (3)$$

On losing a water molecule and undergoing structural arrangement, iron hydroxide forms goethite (FeOOH), a common iron(III) oxide. The fourth reaction involves the oxidation of additional pyrite by ferric ion, Fe^{3+}:

$$FeS_2 + 14Fe^{3+} + 8H_2O \rightarrow 15Fe^{2+} + 2SO_4^{2-} + 16H^+ \quad (4)$$

This is a self-sustaining reaction that perpetuates the pyrite oxidation process. Essentially, once a little portion of pyrite has been oxidised to Fe^{3+}, the oxidation front begins to expand into the unoxidised zones of the ore.

In the Witwatersrand Basin, South Africa, the gold-containing ores also contain a host of other minerals, including a variety of silicates, pyrite (FeS_2), sphalerite (ZnS), uraninite (U_3O_8), gersdorffite (NiAsS), galena (PbS), arsenopyrite (FeAsS) and monazite ((Ce, La, Pr, Nd, Th, Y)PO_4) [5]. The incidence of AMD in this basin results in the leaching of sulphates and elements such as Zn, Ni, As, U, Th and rare earths from these ores [6-9]. These contaminants can undergo a number of reactions that can lead to their transformation, adsorption and desorption along the flow path [10].

This chapter is aimed at demonstrating how these processes can be comprehensively assessed using geochemical modelling, a useful tool that can be used to gain insight into and understanding of geochemical processes. It complements analytical techniques (that are based on laboratory experiments and field data), enabling interpretation of complex situations that conventional methods cannot handle as well as the prediction and anticipation of how geochemical systems would evolve. This has drawn interest in the use of geochemical modelling to predict geochemical processes over the long term, design and optimise remediation strategies, manage environmental impact, conduct risk assessment and identify parameters of importance in geochemical systems [11].

Several geochemical modelling codes have been developed over the years and used to model various scenarios. These include, among others: the Geochemist's Workbench (University of Illinois-Urbana Champaign), Minteq (US EPA), PHREEQC (US Geological Survey), EQ3/6 (Lawrence Livermore National Laboratory), Mineql, Netpath (US Geological Survey), MEDUSA (KTH, Sweden), Aquachem (Schlumberger Water Services) and WHAM (NERC, UK). For this study, the PHREEQC Interactive code was used.

It is common and good practice in modelling to set goals and conceptualise the modelling scenario when conducting simulations. For this work, the goal or aim has been mentioned above. An overview of the various applications of geochemical modelling, using a conceptual model of a polluted abandoned mine site is presented (Figure 1), explaining the key concepts and assumptions behind the simulations using a compartmentalised approach of the site.

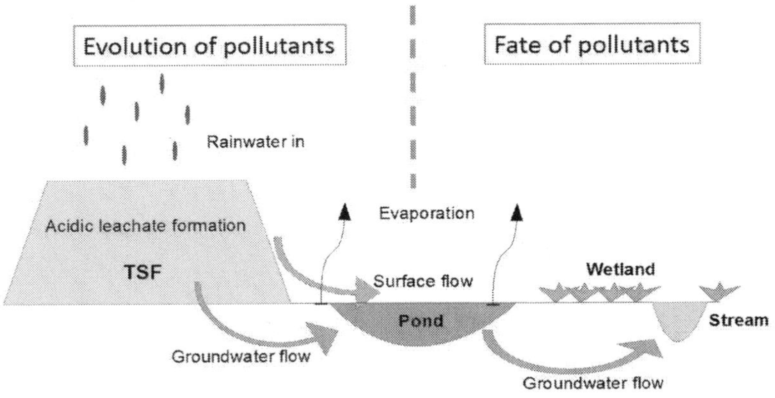

Figure 1. Conceptual model of an abandoned gold mine site (Camden-Smith and Tutu, 2014)

The figure shows a tailings storage facility (TSF) as a source of contaminants owing to AMD generation when the material is contacted with oxygenated rainwater. The leachates released either flow into groundwater or as surface flow, discharging into an adjacent pond. The pond drains into an adjacent wetland due to overflow and groundwater flow and, subsequently, into a natural stream that discharges into an important drainage system to the south of Johannesburg. Some of the important processes along the flow path of released contaminants include changes in elemental speciation; neutralisation; adsorption onto and desorption from hydrous iron oxides (HFOs); and evaporation of leachate solutions, resulting in precipitation of secondary minerals. These processes have been described by different types of models, e.g. speciation-solubility, forward modelling (including neutralisation and adsorption reactions), reaction path models (including evaporation sequences), reactive transport modelling (involving a 1-dimensional flow through system in which contaminants are flushed) [12-15], and inverse modelling (involving finding reactive minerals given an initial and final solution) [13-15].

2. Materials and methods

Solid material was sampled from the oxidised zone of the TSF (a depth of up to 1 m). The distinct orange colour characterises this material, an indication of oxidation of pyrite which is dark greyish in colour. The material clods were pulverised and sieved to < 2mm prior to further treatment.

Due to the unavailability of rainwater during sampling (conducted during the dry season), an artificial rainwater solution was prepared according to the procedure described by Anderson *et al.* [16], but with modification to suit composition of local rainwater in the study area [17]. This was used to leach the tailings at an optimised solid:liquid ratio of 1 g:20 mL (the actual leaching was done on 20 g of solid using 400 mL of artificial rainwater) for 24 h. The leachate was then filtered using a 0.45 μm cellulose nitrate filter paper and geochemical parameters (pH, electrical conductivity, redox potential and temperature) and metals analysed using appropriate probes and inductively coupled plasma optical emission spectroscopy (ICP-OES), respectively.

Three water samples were collected from different points in the pond and mixed to make a composite. Geochemical parameters were determined immediately as previously outlined. The composite sample was then filtered using a 0.45 μm cellulose nitrate filter paper and separated into portions for determination of anions and metals by ion chromatography and ICP-OES, respectively.

Hydrous ferric oxides (HFOs) were extracted and determined according to Breit *et al.* [18]. From the HFO concentrations, the amounts of weak and strong adsorptions sites, Hfo_w and Hfo_s, were determined according to Dzombak and Morel [19].

3. Data treatment and basis for modelling

The above two samples, namely leachate from oxidised tailings and pond water composite will be discussed and utilised in the case studies related to different geochemical models. The analytical data used for modelling are presented in Table 1. For simplicity, only major metals and a selection of minor metals have been presented. The advantages and disadvantages of this will be discussed in the section on speciation modelling.

As mentioned earlier, the PHREEQC geochemical modelling code was used for modelling [13]. This code, developed at the US Geological Survey, is available for free online. It has wide modelling capabilities, with the possibility of using a number of databases, or incorporating new information into its existing databases. The following is worth mentioning regarding databases.

A geochemical model is said to be as good as the quality of the database used to construct it. A database would contain aqueous, mineral and gas phase constituents and reactions. This is usually referred to as the basis. If these constituents, consisting of components and their related species, are not present in the database, they have to be defined in the input file. Their definition includes thermodynamic parameters such as equilibrium constants, entropy and enthalpy values. Thus, the more comprehensive a database, the better the modelling results.

The detailed descriptions related to the different geochemical models are presented for each case study.

Parameter	Pond Water	Rainwater Leachate
	Geochemical parameters	
pH	3.57	3.64
pe	6.4	7.4
Temperature (°C)	18.7	19.8
Conductivity (μS cm^{-1})	1790	700
	Chemical Analyses (mg L^{-1})	
Cl^-	47	<20
SO_4^{2-}	1245	257
Al	33.1	29.1
Ca	195.0	2.4
Co	2.1	0.6
Cu	0.5	0.5
Fe	2.3	0.0
K	10.1	6.7
Mg	96.6	16.3
Mn	27.3	0.8
Na	43.8	0.5
Ni	3.5	1.2
U	0.2	0.2
Zn	4.2	0.2

Table 1. Analysed pond water and artificial rainwater leachate of oxidised tailings

4. Results and discussion

This section presents and discusses the results for the different models obtained from simulating the evolution of the above water samples. Further descriptions of particular models have been included together with the construction protocols of those models.

The input files or scripts have not been included, but information on how these are written or generated can be obtained in the PHREEQC manual and other relevant text [20] that can be downloaded from the USGS website (http://wwwbrr.cr.usgs.gov/projects/GWC_coupled/phreeqc/).

4.1. Speciation-solubility models

The chemical analysis of a sample will generally only provide total concentrations of the elements in the solution. How the elements are related to one another is not easily determined using analytical techniques. Basically, the aim of a speciation model is to predict whether the metal ions are "free" (associated with water molecules only) or has anions in the solution replaced water molecules in the hydration sphere of the metal ion.

Speciation-solubility models are used to define the concentration and activity of stable species in a system and to determine the saturation states of minerals within a system [14]. Most models assume local equilibria in order to solve the equations involved in determining the species spectrum. This assumption is often valid as the reactions that affect the bulk chemistry of natural water bodies occur relatively fast with respect to the time scale of interest [21]. Chemical equilibrium can be solved in two ways, namely by either minimising the Gibbs' free energy of a system or by using mass action equations combined with equilibrium constants. The second method is used by the majority of programming software because of a lack of consistent Gibbs' free energy data [11]. Speciation-solubility models are the simplest geochemical models to define and are the starting models for geochemical reaction models and reactive transport models [14].

For this model, a SOLUTION_SPREAD function was defined in PHREEQC Interactive and the Wateq4f.dat database was used. The output was a long file containing a large amount of information e.g. solution composition following adjustments for charge and electrical balance as well as speciation of elements in solution. Only the results for the latter in pond water will be discussed.

From the input, an output containing the calculated distribution of species in solution was obtained. The results for iron speciation are presented in Table 2.

At a redox potential (pe) value of 6.4 depicted for the solution, iron is expected to exist predominantly as the ferrous valence state. The majority of this iron is in a "free" form, associating through its hydration sphere. The effect of this on activity is quite notable as the activity was found to be significantly lower than molality. The same effect is not observed for the second most abundant species, $FeSO_4$. This neutral species is not expected to have as extensive a hydration sphere as the Fe^{2+}species and as such its molality and activity are approximately equal. The ferric ion is mainly associated with sulphates and hydroxides. The free ferric iron has a much lower activity than molality because of its small hydration sphere and higher charge. There is a large amount of information that can be gained from inspecting the speciation of elements in a solution. A simple speciation-solubility model can prove useful in explaining laboratory experiments. For example, at these pH and pe conditions, a membrane or polymer designed in the hope of capturing ferric ions might be ineffective as only a small percentage of iron exists in the ferric state and secondly, most of the ferric ions exist as a lower charged sulphate species. This could be of particular relevance in the case of uranium capture. In this example, uranium exists predominantly as an uncharged UO_2SO_4 species. An important consideration is the existence of equilibrium between redox elements. In the initial calculation, the valence states of each redox species are dispersed using the experimental pe. The iron speciation presented here is not necessarily in equilibrium with other redox elements. If the defined solutions are to be used for further modelling, thermodynamic pe as calculated by the software will be used. The redox elements will equilibrate with one another and the dispersal of valence states will be different.

Valence	Species	Molality	Activity
	Fe^{+2}	3.04E-05	1.62E-05
	$FeSO_4$	1.16E-05	1.17E-05
	$FeCl^{+}$	2.92E-08	2.48E-08
	$Fe(HSO_4)^{+}$	2.42E-08	2.04E-08
Fe(II)	$Fe(OH)^{+}$	1.38E-11	1.17E-11
	$Fe(OH)_2$	2.10E-19	2.12E-19
	$Fe(OH)_3^{-}$	3.24E-26	2.75E-26
	$Fe(HS)_2$	0.00E+00	0.00E+00
	$Fe(HS)_3^{-}$	0.00E+00	0.00E+00
Total Fe(II) Species		4.21E-05	
	$FeSO_4^{+}$	1.53E-10	1.30E-10
	$Fe(OH)^{2+}$	9.47E-11	4.89E-11
	$Fe(OH)_2^{+}$	5.54E-11	4.71E-11
	$Fe(SO_4)_2^{-}$	1.51E-11	1.27E-11
	Fe^{+3}	1.04E-11	2.98E-12
	$Fe(HSO_4)^{2+}$	1.86E-13	9.46E-14
Fe(III)	$FeCl^{+2}$	1.58E-13	8.16E-14
	$Fe(OH)_3$	1.69E-14	1.70E-14
	$FeCl^{2+}$	5.88E-16	4.97E-16
	$Fe_2(OH)_2^{4+}$	1.27E-18	8.40E-20
	$FeCl_3$	5.48E-20	5.52E-20
	$Fe(OH)_4$	5.24E-20	4.45E-20
	$Fe_3(OH)_4^{5+}$	1.04E-25	1.50E-27
Total Fe (III) species:		3.29E-10	

Table 2. Species distribution of iron in pond water

The final section of the output lists the saturation indices (SI) of the minerals that could be possibly be associated with the solution. The SI of the mineral is the log of the ratio between the ion activity product (IAP) and the solubility constant (K_{sp}) [22]. An SI value of zero indicates that the mineral is in equilibrium with the solution while a negative SI value indicates that the system is undersaturated with the mineral. If the mineral was present, then there would be a possibility that the mineral would dissolve. A positive SI value indicates that the system is supersaturated with respect to that mineral and as such, the solution could precipitate that mineral. There are several reasons as to why a mineral can be supersaturated in solution but not precipitate. These include: kinetic constraints (SI values give no indication of the rate of precipitation reactions); calculated SI values are for pure end members of a mineral series and not reflective of the solid solution composition that would precipitate; errors in analysis and unreliable thermodynamic data [14].

A selection of saturation indices for the pond water is presented in Table 3. According to the indices, alunite, goethite and gypsum are close to equilibrium with the pond water. Pyrite is undersaturated as the activity of the sulphide ion is very low as all of the sulphur is speciated as sulphate or bisulphate ion. Gaseous oxygen is undersaturated. For a gas, the SI is equal to the logarithm of the fugacity. Oxygen was not defined as a species in the initial model as it was not analysed. If the solution was in equilibrium with atmospheric oxygen, then the saturation index would be -0.7. In the next section, the solution will be equilibrated with atmospheric gases.

Name in database	SI	Formula
$Al(OH)_3(a)$	-4.76	$Al(OH)_3$
Alunite	0.93	$KAl_3(SO_4)_2(OH)_6$
Anhydrite	-0.75	$CaSO_4$
Boehmite	-2.57	AlOOH
Epsomite	-3	$MgSO_4{:}7H_2O$
$Fe(OH)_3(a)$	-5.71	$Fe(OH)_3$
Goethite	-0.05	α-FeOOH
Gypsum	-0.51	$CaSO_4{:}2H_2O$
Jarosite-K	-12.8	$KFe_3(SO_4)_2(OH)_6$
Jarosite-Na	-15.93	$NaFe_3(SO_4)_2(OH)_6$
$O_2(g)$	-45.33	O_2
Pyrite	-68.77	FeS_2

Table 3. Selected mineral saturation indices for the pond water sample

Modelled results modelling often reveal more information than analytical results. As indicated earlier, in order to gain a better model of the pond water, atmospheric gases were added as equilibrium phases. Since the pond water samples were obtained from the upper 30 cm of the pond, it is reasonable to assume that they are in equilibrium with oxygen, carbon dioxide and nitrogen. As nitrogen is an inert gas, it was not included as it would not affect the speciation or solubility of the solution. The concentration of gases in the solution was determined by taking into account their atmospheric abundance and using Henry's Law. In order to model the equilibration of the solution at 1 atm with atmospheric gases, the log(fugacity) or saturation index of oxygen was set to -0.7 and carbon dioxide to -3.5. For the pond water, the model predicted that 0.2940 mmol of O_2 and 0.0128 mmol of CO_2 were added to the solution in order to equilibrate the solution with the atmosphere. The addition of oxygen caused the pe to increase to 17.58. This is the thermodynamic pe of the solution. The redox species were then defined according to this pe. At higher pe values, ferric species were the major components. If ferrous and ferric concentrations had been quantified, then it would have been possible to decouple ferrous and ferric iron and they would not speciate according to the defined or calculated pe. It is also possible to define the iron concentration as consisting solely of the ferrous species. This could be a reasonable assumption in cases where the pH of the solution is greater than 4 since, at this pH, precipitation of ferric hydroxide would have removed any ferric species from solution. However, it should be noted that for further reactions, the programme will cause the solution to equilibrate and this will result in either the consumption of oxygen (if defined as an element in the initial solution) or the oxidation of ferrous iron in the presence of a constant oxygen supply (if oxygen is defined as an equilibrium phase). The effect on iron speciation of the addition of atmospheric gases on the initial solution is summarised in Figure 2 and Figure 3. At the experimental pe, iron is predominantly a ferrous species as discussed previously. After equilibration with oxygen and carbon dioxide, the iron was found to be predominantly complexed ferric species. The amount of oxygen gas added to the solution to reach equilibrium was 4.54 mg L^{-1}, while the experimental value obtained on site using a dissolved oxygen probe was 4.6 mg L^{-1}, showing that surface water was indeed equilibrated with atmospheric oxygen. The true nature of iron speciation probably lies between the two models. It should be noted that redox disequilibrium is common in natural waters. As the water equilibrates, both within itself and with the atmosphere, the second model becomes more representative of the system.

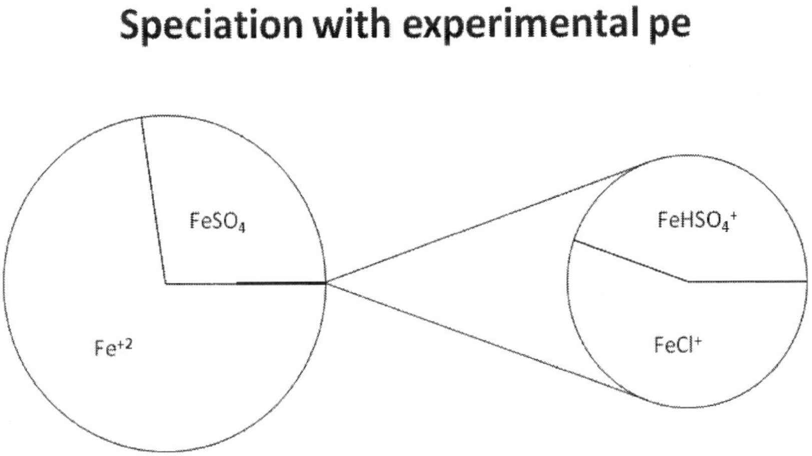

Figure 2. Speciation of iron using experimental pe

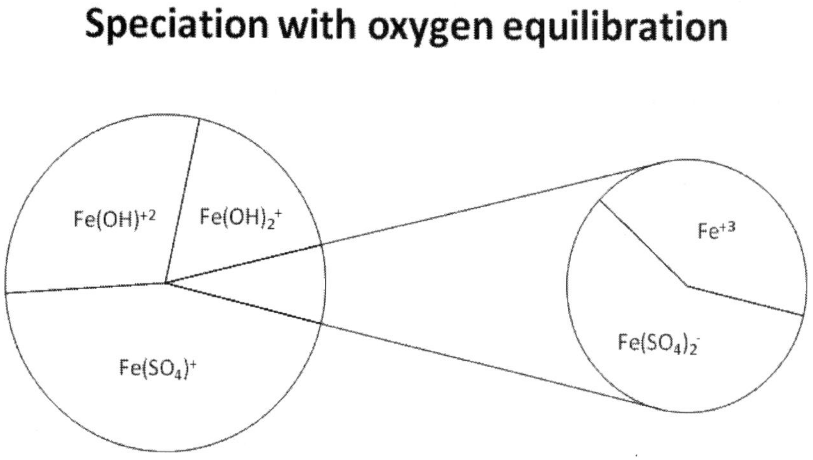

Figure 3. Speciation of iron after equilibration of the solution with atmospheric gases

4.2. Forward models

In forward modelling, the final composition of a solution after a reaction or equilibration is calculated [11]. Thermodynamic models utilise equilibrium constraints and are assumed to be valid for homogeneous reactions in which there is sufficient residence time for equilibrium to be established. Heterogeneous reactions such as the dissolution and precipitation of minerals as well as adsorption are often kinetically controlled and models should be adjusted accordingly. Reaction path models are types of forward models that can track reaction progress in small incremental steps. At each step, the species distribution and saturation indices are calculated and equilibrium is maintained by the dissolution or precipitation of defined minerals [11]. These models are used to describe processes such as the addition of minerals to a water system, the mixing of one solution with another (mixing models) and the removal of minerals or components from a system e.g. evaporation sequences [14].

To demonstrate how forward modelling works, the studied pond water was taken through two remediation processes, namely reaction with calcite ($CaCO_3$) and contact with iron oxide surfaces (i.e. adsorption). These two processes were then modelled simultaneously, simulating reactions that are likely to occur in an environment where the two processes co-exist.

Modelling of the addition of calcite to the pond water was undertaken as follows:

- The pond water was equilibrated with atmospheric gases (oxygen and carbon dioxide)

- Calcite was added as an equilibrium phase with a saturation index of 0.0 (this was to determine the maximum amount of calcite that will dissolve into the pond water)
- The step-wise addition of calcite was carried out, with the solution assessed after each step.
- Minerals that became oversaturated during the addition of calcite were allowed to precipitate.

The final model for the addition of calcite to pond water is presented in Figures 4-10. There is buffering of pH with initial additions of calcite up to just below 250 mg, beyond which the pH increases significantly. The corresponding drop in redox potential (pe) at higher pH values is expected and demonstrated clearly in pe-pH diagrams. The addition of calcite resulted in the precipitation of five minerals, namely diaspore (AlOOH); iron hydroxide ($Fe(OH)_3$); manganite (MnOOH); malachite ($Cu_2(OH)_2CO_3$) and nickel hydroxide ($Ni(OH)_2$). Diaspore was the first mineral to precipitate. It continued to precipitate until all the aluminium in solution was depleted. A similar path was observed when defining $Al(OH)_3$ as the mineral, although diaspore remains supersaturated during the precipitation. Amorphous iron hydroxide precipitated early on in the profile. The iron concentration also decreased to almost complete depletion. Manganite exhibited a similar profile. As the carbonate content and pH increased in the solution, malachite ($Cu_2(OH)_2CO_3$) was predicted to precipitate very late in the profile (just before the saturation of calcite was accomplished). Nickel hydroxide also precipitated during this step. The total concentration of the other defined elements remained unaffected. However, the speciation of elements changed with increasing pH. For example, initially uranium existed predominantly as UO_2SO_4 and UO_2^{2+}species. After the addition of calcite, the carbonate species $UO_2(CO_3)_2^{2-}$and $UO_2(CO_3)_3^{4-}$were found to predominate.

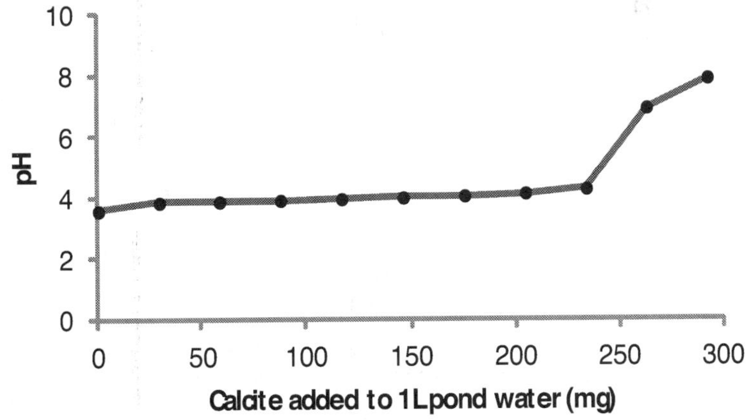

Figure 4. Modelled pH profile during addition of calcite to 1 L pond water

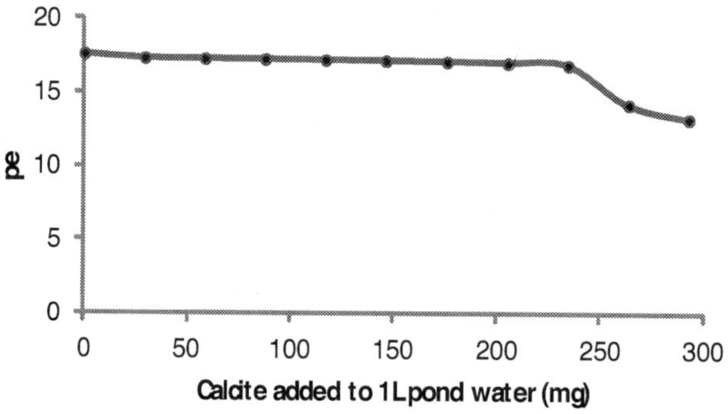

Figure 5. Modelled pe profile during addition of calcite to 1 L pond water

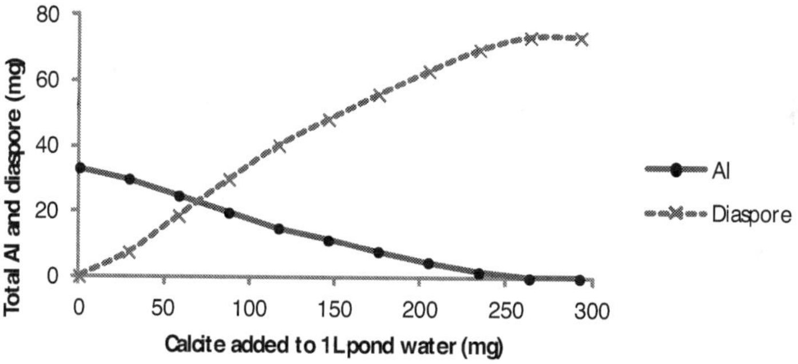

Figure 6. Modelled aluminium profile during addition of calcite to 1 L pond water

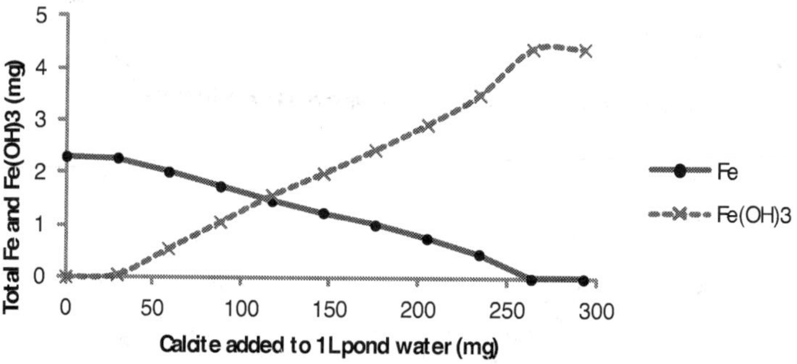

Figure 7. Modelled iron profile during addition of calcite to 1 L pond water

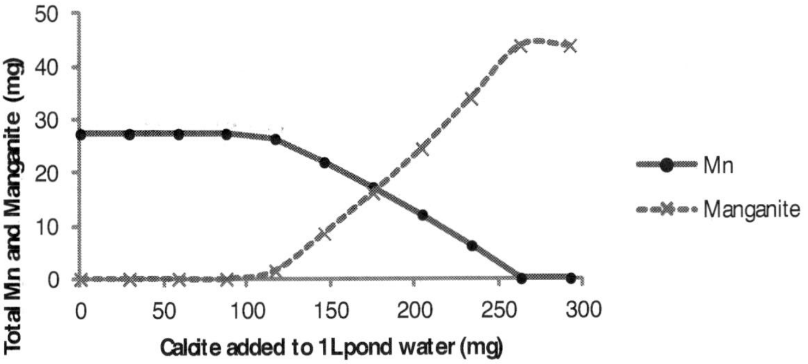

Figure 8. Modelled manganese profile during addition of calcite to 1 L pond water

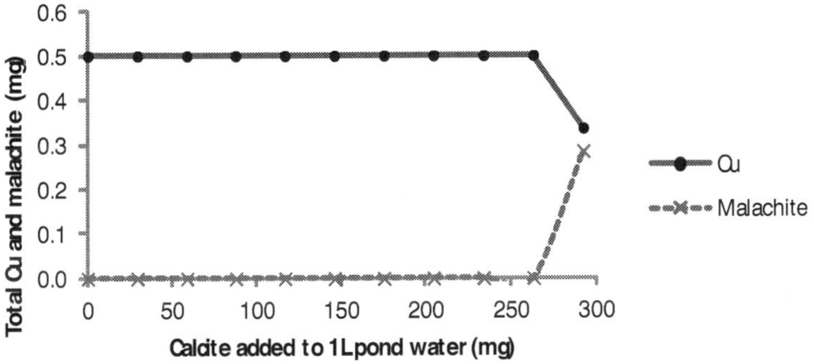

Figure 9. Modelled copper profile during addition of calcite to 1 L pond water

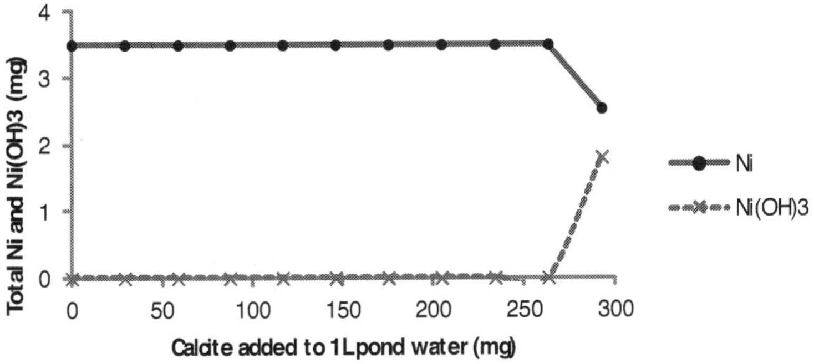

Figure 10. Modelled nickel profile during addition of calcite to 1 L pond water

The addition of ferrihydrite (hydrous ferric oxide (HFO)) to the pond water was modelled. Ions in solution adsorb onto the surface of HFO through strong and weak sites [19, 23]. Essentially, strong sites and weak sites depict sites of high and low energy, respectively. Strong complexation of ions to HFO would occur at strong sites while there would be weak complexation at weak sites. An example of adsorption of Zn^{2+}onto HFO is given below:

$$\equiv FeO^{-} + Zn^{2+} \rightarrow \equiv FeOZn^{+} \tag{5}$$

For every 1 mmol of HFO, a total of 0.2 mmol weak sites and 0.005 mmol strong sites were allocated in the model according to Dzombak and Morel [19] and Appelo *et al.* [23]. The addition of 1 g of HFO to 1 L of pond water equilibrated with oxygen and carbon dioxide was modelled. The following were input parameters for the surface: 0.0112 mol of weak sites, 5.6 x 10^{-5} mol of strong sites and a surface area of 600 m^2 g^{-1}.

The addition of 1 g of HFO increased the pH of the solution to 5.83, largely due to the removal of H^+ ions from solution through adsorption onto HFO. This caused $Al(OH)_3$, diaspore, $Fe(OH)_3$ and manganite to become supersaturated. By setting them as equilibrium phases and allowing them to precipitate, the precipitation of these minerals upon the addition of HFO can be modelled.

The site occupancy of the HFO surface is summarised in Table 4. Approximately, half of the weak sites are occupied by water molecules. Sulphate and hydroxide ions take up the bulk of the remainder of weak sites. Less than 1% are occupied by bivalent oxycations, namely zinc, copper, manganese, nickel. The strong sites are occupied by these oxycations, with the remainder being filled by calcium hydroxide species, hydroxide and uranyl oxide. The sorbed fractions of the metals are summarised in Table 5. This is the percentage reduction in total soluble content of the metal after the addition of 1 g of HFO. Nearly all of the copper and uranium in the pond water has been removed from solution after the addition of HFO. Uranium is sorbed onto the strong sites and copper is split between weak and strong sites. The HFO had no effect on the total concentration of aluminium, chloride, iron, potassium, magnesium and sodium.

Surface complex	Number of sites occupied (mmol)	Percentage of sites
Weak sites		
$Hfo\_wOH_2^{+}$	1.149	51.3
$Hfo\_wSO_4^{-}$	0.589	26.3
$Hfo\_wOHSO_4^{2-}$	0.247	11
Hfo_wOH	0.242	10.8
Hfo_wOZn^{+}	0.004	0.2
Hfo_wOCu^{+}	0.004	0.2
Hfo_wOMn^{+}	0.002	0.1

Surface complex	Number of sites occupied (mmol)	Percentage of sites
Hfo_wONi^{+}	0.002	0.1
Hfo_wO^{-}	0.001	0.1
Strong sites		
Hfo_sOZn^{+}	0.020	35.8
Hfo_sOMn^{+}	0.011	19.9
Hfo_sONi^{+}	0.006	11.6
$Hfo\_sOH_2{}^{+}$	0.006	11.4
Hfo_sOHCa^{2+}	0.006	10.8
Hfo_sOCu^{+}	0.004	7.2
Hfo_sOH	0.001	2.4
$Hfo\_sOUO_2{}^{+}$	0.000	0.9

Table 4. Surface complexes formed after addition of 1 g HFO to 1 L of tailings pond water

Element	Percentage sorbed
Ca	0.1
Cu	98.8
Mn	2.6
Ni	13.5
S	6.4
U	99.9
Zn	37.0

Table 5. Percentage of total elements adsorbed after addition of 1 g HFO to 1 L of pond water

By allowing minerals that become supersaturated during addition of HFO to precipitate, a different model is observed. The pH only increases to 3.94 (due to the removal of hydroxide ions during precipitation). As such, the speciation of metal ions is different to that discussed above. Less than 0.1% of weak sites are occupied by metal ions. Water molecules occupy 52.5% and sulphate ions occupy 47% of weak sites. Water occupies 97% of the strong sites with uranium, calcium and copper occupying approximately 1% each. The models presented are based on thermodynamic principles only, no kinetics have been included. The relative rates of adsorption onto the HFO surface and of precipitation of minerals will determine the model that better represents the chemistry of the system.

A model combining the addition of calcite with the addition of HFO can be built. These two reactions are important in situations where water remediation strategies are being considered, for example, the use of a permeable reactive barrier. For this combined model, pond water was equilibrated with oxygen and carbon dioxide and reacted with calcite until saturation of calcite was achieved. Minerals that became supersaturated were allowed to precipitate and the resulting solution brought into contact with 1 g HFO. The final predicted water quality for this model is summarised in Table 6. The percentage reduction shown is with respect to the original solution. The pH and pe of the final modelled solution were 7.32 and 13.8, respectively. The addition of calcite increased the calcium concentration in the final solution, hence a negative value. There was a major reduction in the concentrations of aluminium, copper, iron, manganese, nickel, uranium and zinc. Chloride, potassium, magnesium, sodium and sulphate values were not significantly reduced. The strong sites of the HFO had sorbed zinc (41%), calcium (35%), nickel (21%) and copper (1%). Weak sites were dominated by hydroxides and sulphate with 12% occupation for magnesium and smaller amounts of zinc (2%), calcium (1%) and nickel (1%).

Parameter	Total concentrations (mg/L)	Percentage reduction
Al	0.0	100
Ca	310	-58.9
Cl	47	0.0
Cu	0.0	99.97
Fe	0.0	99.97
K	10.1	0.0
Mg	90.4	6.6
Mn	0.0	100.0
Na	43.9	0.0
Ni	0.3	90.4
SO_4^{-2}	1164	2.5
U	0.0	99.5
Zn	0.1	97.6

Table 6. Final quality of pond water after addition of 1 g HFO to 1 L initial pond water

Types of reaction path models include: titration, buffering, flush and kinetic reaction path models [11]. The previous example of calcite dissolution is a typical case of a titration or buffering/neutralisation model. Evaporation sequences are some of the most interesting models as they tend to reveal a wide range of possible minerals, thus complementing analytical techniques such as X-ray diffraction (XRD) that may come short in determining some minerals.

An evaporation model of the pond water is presented as Figure 11. The model was prepared assuming equilibration with the atmospheric gases, O_2 and CO_2. As such, the iron is speciated from the beginning of the calculation as ferric iron. The ferric species is insoluble and was allowed to precipitate at the beginning of the reaction as jarosite ($KFe^{3+}{}_3(OH)_6(SO_4)_2$). Goethite ($\alpha$-FeOOH) could also have selected as an initial phase. The model was created by removing water incrementally using the REACTION function and allowing minerals to precipitate by setting them as EQUILIBRIUM_PHASES. The first model allowed for the dissolution of precipitated minerals as the pH changed (Figure 11). This would be relevant for a closed system, for example an evaporation experiment occurring in a beaker (Figure 12). The water would remain in contact with the precipitated minerals and as the pH decreases during evaporation, the minerals become unsaturated in solution (saturation indices decrease to below zero) and the precipitated minerals dissolve. For a more environmentally relevant evaporation profile in an open system, as in the case of the pond water, the subsequent dissolution would not be allowed because the minerals that precipitate on the edge of the pond will no longer be in contact with the remaining water (Figure 12). Therefore, even though the solution becomes unsaturated with respect to the mineral, no resolubilisation occurs because the minerals have left the system. In both cases, alunite ($KAl_3(SO_4)_2(OH)_6$) and jarosite were supersaturated at the start of the evaporation profile. This is not ideal for an evaporation model as we are assuming that the analysed solution is in thermodynamic equilibrium, meaning that a solution in equilibrium would have already precipitated saturated minerals. Minerals can be supersaturated in natural waters as their precipitation could be kinetically hindered. In general, it is preferable that minerals be saturated during evaporation as in the case of gypsum ($CaSO_4.2H_2O$) as shown in the profiles in Figures 11 and 12. Notwithstanding, the models are still useful as an indication of the type of minerals that can be precipitated from the water. The minerals presented are pure end members and in nature would precipitate as solid solution minerals, thereby removing some of the transition metals from solution.

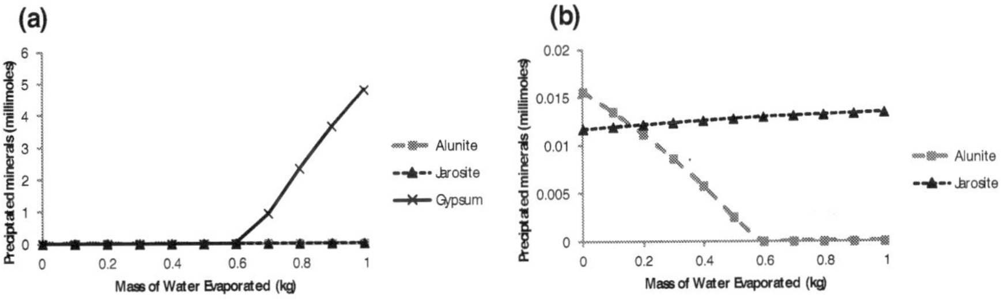

Figure 11. Closed system evaporation model of water (a) Precipitation of gypsum and (b) zoomed in precipitation of jarosite and precipitation and redissolution of alunite.

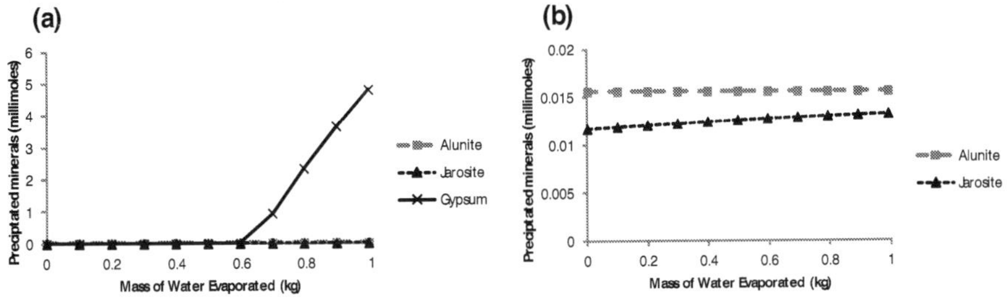

Figure 12. Open system evaporation model of water (a) Precipitation of gypsum and (b) zoomed in precipitation of jarosite and precipitation and redissolution of alunite.

4.3. Inverse models

Inverse modelling is also known as mass balance modelling. Given the composition of two water systems along the same flow path and the mineralogy of the rock through which the water flows, inverse modelling can be used to provide a set of possible reactions that occurred to transform the first solution into the second [24]. Mass balance equations are utilised and neither thermodynamic properties nor kinetic restraints are considered [14].

Inverse modelling of the rainwater leachate is presented in Table 7. The initial solution was the blank experiment that was subjected to the same leaching and analytical methods as the leachate but did not contain any solid material. Table 7 summarises the phase transfers from the initial solution into and from the leachate. Phases were defined based on metal fractionation information gained from sequential extractions of the tailings material. It revealed that most of the metals discussed in this chapter exist as readily soluble phases in the oxidised tailings except for uranium which was associated with a reducible phase. The reducible phases are regarded as iron and manganese oxyhydroxides (FeOOH and MnOOH). It is possible that uranium does not exist in the material as a well-defined uranyl hydroxide, but is strongly adsorbed onto the surface of the iron hydroxide. For the sake of simplicity, however, uranium has been allocated as its own phase. The defined phases were chalcanthite ($CuSO_4.5H_2O$), epsomite ($MgSO_4$), $MnSO_4$, morenosite ($NiSO_4.7H_2O$), β-$UO_2(OH)_2$, gypsum, $Al_2(SO_4)_3$:$6H_2O$, $ZnSO_4$:H_2O, $Al(OH)_3$(a), $Fe(OH)_3$(a) and melanterite ($FeSO_4.7H_2O$). The phase data for $Al_2(SO_4)_3$:$6H_2O$ was not available in the Wateq4f.dat database and so had to be imported from the llnl.dat database. For simplicity, chloride (Cl^-) and nitrate (NO_3^-) were excluded from the model. Including these species required the definition of more phases, leading to a large number of potential models. No phase including potassium was defined. Model 1 (Table 7) is the simplest model predicted by the program. It contains the minimum number of phases. In the input file, the uncertainty parameter for each solution must be defined. In order to allow for the convergence of the initial solution, a large uncertainty was allocated for the concentrations. This is often required when attempting to apply inverse models to dilute solutions. Model 1 was generated by allowing for the adjustment of calcium, sodium and sulphate values. As such, there were no calcium or sodium containing phases in the model. Model 2 (Table 7) had more phases. Only the sulphate value was adjusted by the program in order to generate the model. The sulphate values in the initial rainwater and leachate solutions were adjusted by 3% and 5%, respectively. As illustrated in this model, the software will use minerals as defined by the user. The iron concentration in both solutions was below the detection limit, but, despite this, melanterite and amorphous iron hydroxide were defined as phases. Model 2 predicts the dissolution of the ferrous sulphate mineral followed by the precipitation of ferric hydroxide. The precipitation of amorphous aluminium hydroxide in Model 1 (and to an extent in Model 2) assists in lowering the pH of the initial solution to equate to that in the final solution. Forward modelling could be undertaken in order to look at the validity of each model. This would involve dissolving the predicted minerals in rainwater to see if this would yield the leachate solution.

Phase	Amount (moles)	Mass (mg)	Action	Formula
Model 1				
Chalcanthite	7.70×10^{-6}	1.9	Dissolve	$CuSO_4.5H_2O$
Epsomite	6.65×10^{-4}	163.9	Dissolve	$MgSO_4.7H_2O$
$MnSO_4$	1.46×10^{-5}	2.2	Dissolve	$MnSO_4$
Morenosite	2.05×10^{-5}	5.7	Dissolve	$NiSO_4.7H_2O$
β -$UO_2(OH)_2$	8.41×10^{-7}	0.3	Dissolve	$UO_2(OH)_2$
$Al_2(SO_4)_3.6H_2O$	5.75×10^{-4}	258.9	Dissolve	$Al_2(SO_4)_3.6H_2O$
$ZnSO_4.H_2O$	3.06×10^{-6}	0.5	Dissolve	$ZnSO_4.H_2O$
$Al(OH)_3$(a)	7.23×10^{-5}	5.6	Precipitate	$Al(OH)_3$
Model 2				
Chalcanthite	7.70×10^{-6}	1.9	Dissolve	$CuSO_4.5H_2O$
Epsomite	6.65×10^{-4}	163.9	Dissolve	$MgSO_4.7H_2O$
$MnSO_4$	1.46×10^{-5}	2.2	Dissolve	$MnSO_4$
Morenosite	2.05×10^{-5}	5.7	Dissolve	$NiSO_4.7H_2O$
β-$UO_2(OH)_2$	8.41×10^{-7}	0.3	Dissolve	$UO_2(OH)_2$
Gypsum	3.39×10^{-5}	5.8	Dissolve	$CaSO_4.2H_2O$
$Al_2(SO_4)_3.6H_2O$	5.75×10^{-4}	258.9	Dissolve	$Al_2(SO_4)_3.6H_2O$
$ZnSO_4.H_2O$	3.06×10^{-6}	0.5	Dissolve	$ZnSO_4.H_2O$
$Fe(OH)_3$(a)	1.01×10^{-10}	1.1×10^{-5}	Precipitate	$Fe(OH)_3$
Melanterite	1.01×10^{-10}	2.8×10^{-5}	Dissolve	$FeSO_4.7H_2O$
Thenardite	3.87×10^{-6}	0.5	Dissolve	Na_2SO_4
$Al(OH)_3$(a)	7.23×10^{-5}	5.6	Precipitate	$Al(OH)_3$

Table 7. Inverse modelling of rainwater leachate

4.4. Reactive transport models

Modelling the effect of calcite and HFO on pond water chemistry as it leaches through a medium containing these best exemplifies a reactive transport model. For this purpose, a permeable reactive barrier (PRB) containing calcite and HFO was considered (conceptualized in Figure 13), with initial conditions defined as follows:

- Division of different areas into compartments as follows: zone 1 containing pond water; zone 2 containing 9 cells that had the same amount of calcite and HFO and zone 3 containing 3 cells that have only rainwater in them.
- Zone 2 above is the PRB barrier and was "filled" with rainwater that was already equilibrated with calcite.

• Zone 3 did not contain any calcite nor HFO.

The simulation was such that a band of pond water was allowed to flow through the PRB, with the final predicted water quality represented in zone 3. The assumption was that pond water flowed through the zones followed by rainwater.

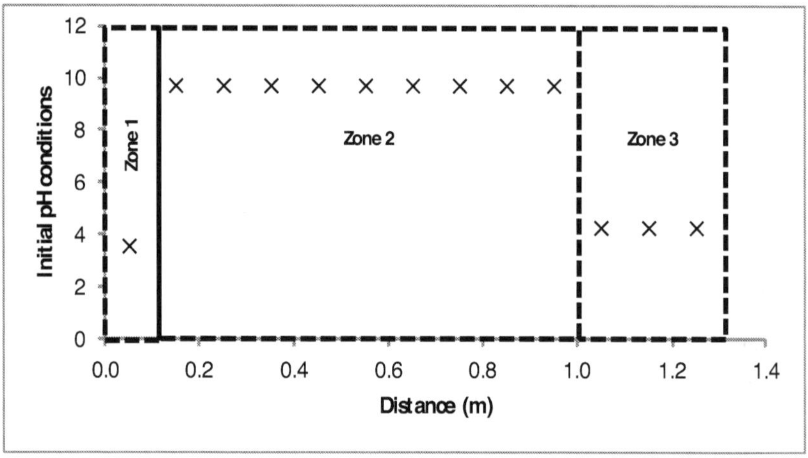

Figure 13. Conceptual model and initial conditions of reactive barrier

Simulation of the transport of iron through the zones gave the results presented in Figure 14. The "steps" are time intervals that help to track the changes along the flow path. Iron was found to be transported conservatively, with concentrations remaining just above 4 x 10^{-5} mol L^{-1}.This is as a result of no precipitation minerals being defined. The speciation of iron changed with change in pH and pe, but the total concentration remained constant.

The results for transportation of manganese are presented in Figure 15. Manganese was found to bind to the surface of HFO. Initially, there is a large decrease in manganese concentration with most of it being captured by HFO sites at the front of the barrier (as shown by step 4 in Figure 16). However, as rainwater flowed through after the pond water, it was shown to leach off manganese from the adsorptive sites, resulting in increased concentrations in subsequent solutions (as shown by step 8 in Figure 15). Thereafter, the manganese concentration is spread out in the barrier as it adsorbs onto HFO and is then slowly leached off (step 16 and step 20 in Figure 15). The effects of strong and weak HFO sites on the binding of manganese are shown in Figures 16 and 17, respectively. Weak sites had higher initial manganese (a magnitude higher than for strong sites) adsorbed onto them and this decreased to lower than for strong sites following flushing with rainwater. Strong sites showed a high resistance to desorption of manganese adsorbed onto them as shown by the almost constant adsorbed concentration (Figure 16) while weak sites, on the other hand, showed a low resistance to desorption of manganese by rainwater (Figure 17). The desorption steps showed that the adsorbed concen-

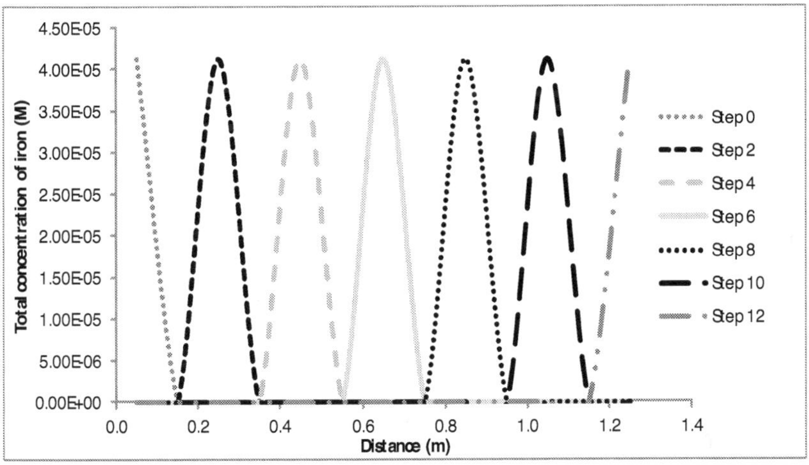

Figure 14. Predicted transportation of iron through the barrier

trations decreased over time, evidence that manganese was lost from solid surface to the water column. This increased manganese content in water is the sustained content observed in Figure 15 as explained before. The observed decrease in manganese concentration on weak sites could be attributed to competition for these sites by calcium and magnesium contained in rainwater (owing to contact with calcite).

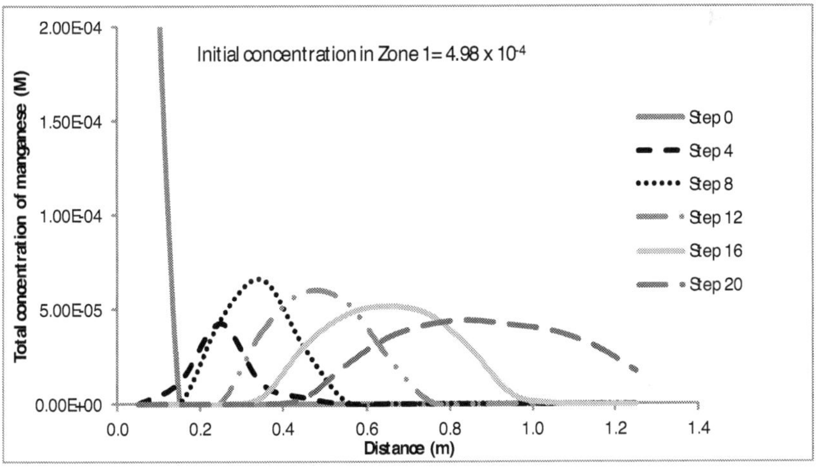

Figure 15. Predicted transport of manganese through barrier

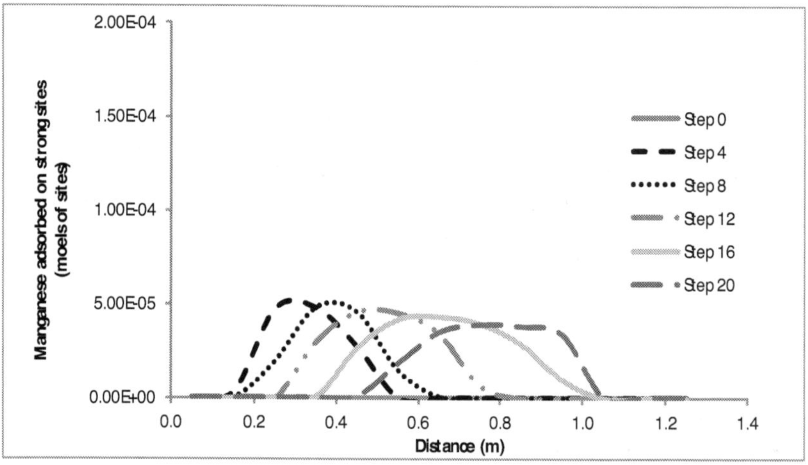

Figure 16. Manganese adsorption on strong sites

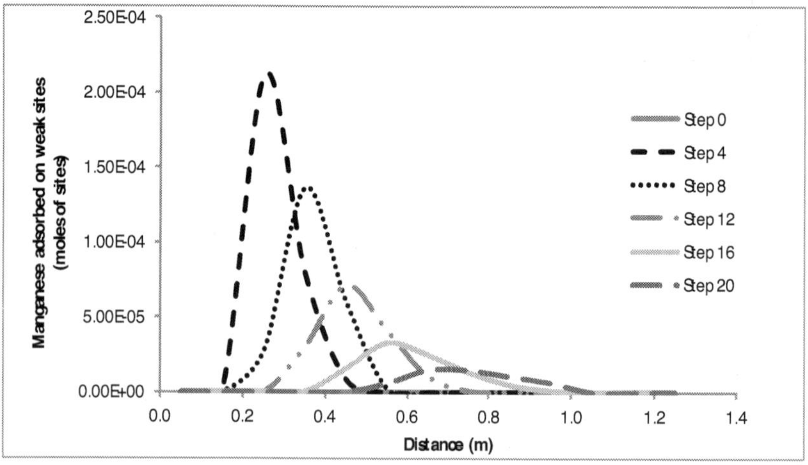

Figure 17. Manganese adsorption on weak sites

5. Limitations to geochemical modelling

Geochemical modelling is an important tool for predicting reactions. However, there are some limitations to its capability that should be considered. For instance, the quality of the data used for modelling influences the quality of the models obtained. While some manipulations can be done on the data during the modelling process, the best practice requires that analytical data be as accurate and relevant as possible. Another limitation is the availability and suitability of databases. This is encountered when modelling of solutions of high ionic strengths, e.g. brines, has to be conducted. The commonly used Debye-Hückel and Davies expressions have shortcomings in handling such solutions as they are suitable for lower ionic strengths [13], and the more robust Pitzer expression [25] lacks reliable literature of most data, particularly for redox sensitive species. The other challenge is that of a general lack of literature containing kinetic data for a number of important mineral reactions.

6. Conclusion

Geochemical modelling has been shown to be a useful tool in making predictions of contaminant behaviour in the aqueous environment. This has been demonstrated by modelling the same water samples using various models. While it may have some limitations, geochemical modelling remains a robust method that gives a better understanding of contaminants and that can be used for decision making such as in proposing appropriate remediation strategies for contaminated water, predicting potential impacts of contaminants on the environment and for general risk assessment. It can be used to complement analytical techniques, revealing important information where the latter tend to fail.

Acknowledgements

The authors would like to acknowledge the National Research Foundation of South Africa, the National Academy of Science (through the PEER Program) and the University of the Witwatersrand for the financial support towards this work.

Author details

Bronwyn Camden-Smith[1], Raymond H. Johnson[2], Peter Camden-Smith[3] and Hlanganani Tutu[1*]

*Address all correspondence to: hlanganani.tutu@wits.ac.za

1 Molecular Sciences Institute, School of Chemistry, University of the Witwatersrand, Johannesburg, South Africa

2 SM Stoller Corporation, Grand Junction, CO, USA

3 Camden Geoserve, Boksburg, South Africa

References

[1] Singer, P.C. and Stumm W. (1970). Acid mine drainage: rate determining step, *Science* 167, 1121-1123.

[2] Stumm, W. and Morgan, J.J. (1981). Aquatic chemistry, An Introduction Emphasising Chemical Equilibria in Natural Waters, 2nd edition, Wiley, New York.

[3] Nordstrom, D.K., Jenne, E.A. and Ball, J.W. (1979). Redox equilibria of iron in acid mine waters. In: Jenne, E.A. (Ed.): Chemical modeling in aqueous systems. Am. Chem. Soc. Symp. Washington, D.C., Series 93, 51-79.

[4] Blowes, D.W., Ptacek, C.J., Benner, S. G., Waybrant, K.R., Bain, J.G. (1998). Permeable reactive barriers for the treatment of mine tailings drainage water. In: Proceedings of the International Conference and Workshop on Uranium Mining and Hydrogeology, Freiberg, Germany, Vol. 2, 113-119.

[5] Feather, C.E. and Koen, G.M. (1975). The mineralogy of the Witwatersrand Reefs, *Minerals Science Engineering*, 7, 189-224.

[6] Rosner, T. and van Schalkwyk, A. (2000). The environmental impact of gold mine tailings footprints in the Johannesburg region, South Africa, *Bull. Eng. Geol. Env.*, Vol. 122, 137-148.

[7] Naicker, K., Cukrowska, E. and McCarthy, T. S. (2002). Acid mine drainage arising from gold mining activities in Johannesburg, South Africa and environs, *Journal of Environmental Pollution* 122, 29-40.

[8] Mphephu, N.F. (2004). Geotechnical environmental evaluation of mining impacts on the Central Rand. PhD Thesis submitted to the University of the Witwatersrand, Johannesburg.

[9] Tutu, H., McCarthy, T.S. and Cukrowska, E.M. (2008). The chemical characteristics of acid mine drainage with particular reference to sources, distribution and remediation: the Witwatersrand Basin, South Africa, as a case study, *Applied Geochemistry* 23: 3666-3684.

[10] Camden-Smith, B.P.C. and Tutu, H. (2014). Geochemical modelling of the evolution and fate of metal pollutants. *Water Science and Technology* 69 (5): 1108-14.

[11] Crawford, J. (1999). Geochemical Modelling – A Review of Current Capabilities and Future Directions. Swedish Environmental Protection Agency, Report 262.

[12] Bethke, C.M. (1996). Geochemical reaction modelling – Concepts and applications, Oxford University Press, New York.

[13] Parkhurst, D.L. and Appelo, C.A.J. (1999). User's guide to PHREEQC (Version2)—A computer program for speciation, batch-reaction, one-dimensional transport, and in-

verse geochemical calculations. U.S. Geological Survey Water-Resources Investigations Report 99-4259.

[14] Zhu, C. and Anderson, G. (2002). Environmental Applications of Geochemical Modeling. The Press Syndicate of the University of Cambridge, Cambridge, United Kingdom.

[15] Šráček, O., Černík, M. and Vencelides, Z. (2013). Applications of Geochemical and Reactive Transport Modeling in Hydrogeology, 1st edition, Palacký University, Olomouc, Czech Republic. ISBN 978-80-244-3781-1

[16] Anderson, P., Davidson, C.M., Duncan, A.L., Littlejohn, D., Urea, A.M. and Garden, L.M. (2000). Column leaching and sorption experiments to assess the mobility of potentially toxic elements in industrially contaminated land. *J. Environ. Monit.* 2: 234-239.

[17] SA Weather Bureau (1998). Climate of South Africa. Weather Bureau Publication WB40, Department of Environmental Affairs and Tourism, Pretoria, 475.

[18] Breit, G.N.; Tuttle, M.L.W.; Cozzarelli, I.M.; Berry, C.J.; Christenson, S.C.; Jaeschke, J.B. (2008). Results of the Chemical and Isotopic Analyses of Sediment and Ground Water from Alluvium of the Canadian River Near a Closed Municipal Landfill, Norman, Oklahoma, Part 2. USGS Open-File Report: 2008-1134

[19] Dzombak, D.A. and Morel, F.M.M. (1990). Surface complexation modeling. Hydrous ferric oxide, John Wiley & Sons, New York.

[20] Parkhurst, D.L. and Appelo, C.A.J. (2013). Description of input and examples for PHREEQC version 3--A computer program for speciation, batch-reaction, one-dimensional transport, and inverse geochemical calculations: U.S. Geological Survey Techniques and Methods, book 6, chap. A43, 497 p. Available at http://pubs.usgs.gov/tm/06/a43/

[21] van der Lee, J. and De Windt, L. (2001). Present state and future directions of modeling of geochemistry in hydrogeological systems. Journal of Contaminant Hydrology, 47 (2-4), 265-2852.

[22] Nordstrom, D. and Munoz, J. (1986). Geochemical Thermodynamics. Palo Alto: Blackwell Scientific Publications.

[23] Appelo, C., van der Weiden, M., Tournassat, C., & Charlet, L. (2002). Surface complexation of ferrous iron and carbonate on ferrihydrite and the mobilisation of arsenic. Environmental Science and Technology, 36 (14), 3096-3103.

[24] Nordstrom, D. (2007). 5.02-Modeling Low-Temperature Geochemical Processes. In H. Holland, & K. Turekian, Treatise on Geochemistry (pp. 1-38). Pargamon: Oxford.

[25] Pitzer K.S. (1979). Theory-Ion interaction approach: in R.M. Pytkowitcz, ed., *Activity Coefficients in Electrolyte Solutions*, vol. 1, CRC Press, Inc., Boca Raton, Florida, pp. 157-208.

Chapter 3

Quality of Water Resources in Malaysia

Yuk Feng Huang, Shin Ying Ang, Khia Min Lee and Teang Shui Lee

Additional information is available at the end of the chapter

http://dx.doi.org/10.5772/58969

1. Introduction

The rapid urbanisation and growth of the population has led to both the ever increasing demand forwater consumption and in tandem the levels of water pollution in Malaysia. Rapid development has produced great amounts of human wastes, including domestic, industrial, commercial and transportation wastes which inevitably ends up in the water bodies. A large number of rivers are so polluted that in some, the consequences are to the extent that the rivers cannot be rehabilitate. Consequently, access to clean and safe water supply has become a tremendous challenge for the water authorities to surmount. Recognizing this, conservative water quality monitoring programs and sustainable use of water are promoted. The major pollutants in Malaysian's rivers and lakes are Biochemical Oxygen Demand (BOD), Ammoniacal Nitrogen (NH3-N) and Suspended Solids (SS). High BOD is contributed largely by untreated or partially treated sewage from manufacturing andagro-based industries. The main sources of NH3-N are domestic sewage, livestock farming and other liquid organic waste products, whilst the sources for SS are mostly earthworks and land clearing activities which the removal is generally achieved through the use of sedimentation and/or water filters. The pollution in rivers and lakes has made it necessary for the water providers to take measures toward securing a stable supply of tap water and supplying potable water. On the other hand, the main contaminants of the marine waters in the country are mainly suspended solids, Escherichia coli, and oil and grease, whilst for groundwater are solid waste landfills, radioactive landfill, etc.

2. River water

River water quality and pollution control need to be addressed urgently since 98 percent of the total water use originates from the rivers. 70% of the water resources in the country are for the agricultural industry.As river water pollution increases, concentrations of the existing pollutants increase. Consequently, it increases water 'quantity scarcity' since good quality water available for use decreases and higher water treatment costs due to the presence of new pollutants. Moreover, the ecological health of the water bodies and the surrounding ecosystems degrade, affecting aquatic lives and habitat, and recreational activities.

2.1. River water quality and status

The Department of Environment (DOE) uses Water Quality Index (WQI) and National Water Quality Standards for Malaysia (NWQS) to evaluate the status of the river water quality [1]. The WQI introduced by DOE is being practiced in Malaysia for about 25 years and serves as the basis for the assessment of environment water quality, while NWQS classifies the beneficial uses of the watercourse based on WQI. In 2012, nine rivers within the Klang River Basin under River of Life Project were added to the national river water quality monitoring programme. The river water quality was assessed based on a total of 5,083 samples taken from a total of 473 rivers. Out of 473 rivers monitored, 278 (59 percent) were found to be clean, 161 (34 percent) slightly polluted and 34 (7 percent) polluted [1]. Figure 1 shows the river water quality trend for 2005-2012.

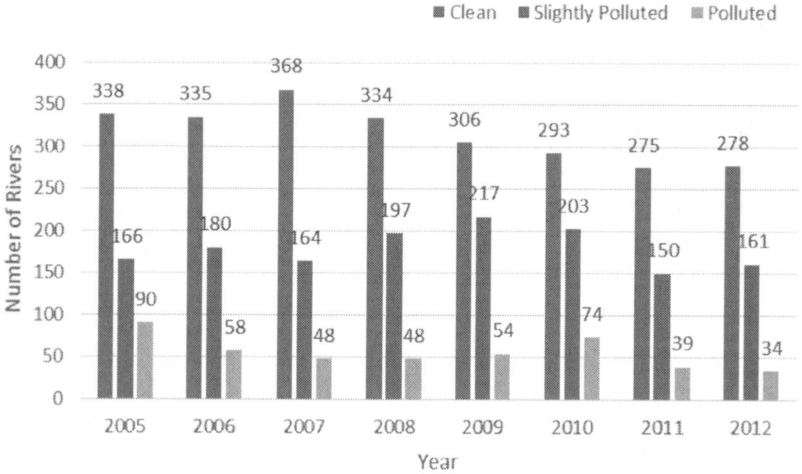

Figure 1. River water quality trend for 2005-2012 [1]

In 2012, 34 rivers were categorized as being polluted, as shown in Figure 1. Out of these, 19 rivers were classified as Class III, 14 rivers as Class IV and one river as Class V. Classification of level pollution by individual pollutant follows the standard set by DOE, as shown in Table 1. Construction activities such as earthworks and land clearing appear to be the main contributor for the sources of SS, whilst the sources for BOD and NH3-N were mostly from agro-based industries and livestock farming, respectively. Pollution of river by untreated or partially treated sewage was also indicated in term of BOD and NH3-N. Besides pollution from organic

http://dx.doi.org/10.5772/58969

pollutant, inorganic pollutant especially heavy metals also another crucial contribution. Mercury (Hg), Arsenic (As), Cadmium (Cd), Chromium (Cr), Plumbum (Pb), and Zinc (Zn) were analyzed. All Pb, Cd and Zn data were within the Class IIB limits of the NWQS. Meanwhile, 99.97 percent of the data for Cr were within the Class IIB limits of the NWQS followed by Hg (99.96 percent) and As (99.68 percent) [1].

Parameter	Unit	Class				
		I	II	III	IV	V
Ammoniacal Nitrogen	mg/l	< 0.1	- 0.3	0.3 - 0.9	0.9 - 2.7	> 2.7
Biochemical Oxygen Demand	mg/l	< 1	1– 3	3 - 6	6 – 12	> 12
Chemical Oxygen Demand	mg/l	< 10	10 – 25	25 – 50	50 – 100	> 100
Dissolved Oxygen	mg/l	> 7	5 – 7	3 - 5	1 - 3	< 1
pH	-	> 7	6 - 7	5 - 6	< 5	> 5
Total Suspended Solid	mg/l	< 25	25 - 50	50 - 150	150 - 300	> 300
WQI	-	< 92.7	76.5 - 92.7	51.9 - 76.5	31.0 - 51.9	> 31.0

Table 1. DOE Water Quality Index Classification [1]

According to the National Water Quality Standards, surface water quality could be gradually improved/upgraded to a better water class based on the standard values of 72 parameters in 6 water use classes. Tables 2 and 3 illustrate the water classes and uses and the DOE Water Quality Classification based on WQI, respectively.

Class	Uses
Class I	Conservation of natural environment. Water Supply I - Practically no treatment necessary. Fishery I - Very sensitive aquatic species
Class IIA Class IIB	Water Supply II - Conventional treatment. Fishery II - Sensitive aquatic species. Recreational use body contact.
Class III	Water Supply III - Extensive treatment required. Fishery III – Commonof economic value and tolerant species;livestock drinking.
Class IV	Irrigation
Class V	None of the above

Table 2. Water Classes and Uses [1]

Sub Index & Water Quality Index	Index Range		
	Clean	Slightly Polluted	Polluted
Biochemical Oxygen Demand(BOD)	91 - 100	80 - 90	0 - 79
Ammoniacal Nitrogen(NH3-N)	92 - 100	71 - 91	0 - 70
Suspended Solids(SS)	76 - 100	70 - 75	0 - 69
Water Quality Index(WQI)	81 - 100	60 - 80	0 - 59

Table 3. DOE Water Quality Classification based on Water Quality Index [1]

2.2. River water pollution

The sources of water pollution can be categorized as point sources and non-point sources. Point sources of pollution are referred to the sources with discharges entering water body at specific locations such as discharges from industries, sewage treatment plants and animal farms, while non-point sources do not have specific discharge point such as surface run-off.In 2012, 1,662,329 water pollution point sources were identified, which comprised of 4,595 manufacturing industries, 9,883 sewage treatment plants (not including individual and communal septic tanks), 754 animal farm (pig farming), 508 agro-based industries, 865 wet markets and 192,710 food services establishments[1].

According to [1], there were three parameters of pollutant loading that have significant impact on the river quality, which include BOD, SS and NH3-N. The estimated BOD load of 848 tons per day in 2012 had decreased by 39 percent compared to 2011 (1,394 tons per day). On the other hand, the estimated SS load was 1383 tons per day, while the estimatedNH3-N load was 232 tons per day, in 2012. Based on the monitoring results in 2012by DOE [1], in terms of river basin basis, Klang River Basin received the highest BOD Load (142 tons per day), followed by Perak River Basin (State of Perak) 114 tons per day, Sarawak River Basin (State of Sarawak) 30 tons per day, Jawi River Basin (State of Pulau Pinang) 26 tons per dayand Muar River Basin (State of Johor) 24 tons per day. Klang River Basin also received the highest SS Load (360 tons per day) and NH3-N load (37 tons per day) among the river basins in Malaysia.

3. Lakes and reservoirs

A lake is an enclosed body of water (usually freshwater) with considerable size, totally surrounded by land with no direct access to sea except with a river or stream that feeds or drains the lake [3, 4]. Lakes are considered lentic systems or standing water compared to rivers which are lotic systems or flowing water [5]. Generally, lakes have in-flowing and out-flowing water which makes them complex [5]. Lakes have 3 main characteristics which make them so unique: long water retention time and complex response dynamics and integrating nature of water body [6]. Lake may occur anywhere within a river basin as natural or man-made lakes [4, 7]. Natural lakes consist of wetlands, marshes, estuarine lakes, or naturally occurring lakes

http://dx.doi.org/10.5772/58969

whereas man-made lakes can be referred as reservoirs, retention pond, ex-mining pond, recreational lakes [8] or as urban landscape elements. Generally, lakes are important source of freshwater which account only the very small part of around 0.01 percent of the global amount of water [7].

State	Number	Area (km^2)	Volume (Mm^3)
Perlis	2	13.33	40.00
Kedah	7	95.03	1,637.76
Perak	11	284.68	6,766.50
Selangor	15	11.38	511.32
Pahang	10	94.69	355.71
Kelantan	4	11.34	76.80
Johor	13	84.22	940.02
Labuan	3	0.50	5.40
Melaka	4	8.75	81.30
Negeri Sembilan	5	2.25	182.33
Penang	4	0.94	47.20
Sabah	5	1.81	29.61
Sarawak	4	97.08	6,080.00
Terengganu	2	370.80	13,600.00
Putrajaya	1	7.50	45.00
Total	**90**	**1,094.89**	**30,398.95**

Table 4. Lakes and reservoirs based on states in Malaysia [8].

Lakes in Malaysia, natural or artificial, have multiple functions. Almost 90 percent of the nation's water supply comes from the lakes and reservoirs [8]. Lakes and reservoirs serves as the source of water for domestic, industrial and agriculture; hydroelectric power generation; flood mitigation, navigation and recreation [8, 9]. They are also home to a variety of biological species and freshwater fish industry [9], i.e. the Bukit Merah Lake in Perak is believed to be the only origin of its natural habitat for the exotic species, Malayan Gold Arowana, *Scleropages formosus* where this fish could cost up to USD 15,000 – USD 25,000 per fish in the market for a full grown adult [10]. Natural lakes in Malaysia include wetlands like Bera Lake and Chini Lake in Pahang, and Dayang Bunting Lake in Langkawi, in addition to ox-bow lakes such as Paya Bungor in Pahang and Logan Bunut in Sarawak [8, 9]. Among the artificial lakes or reservoirs include Kenyir Lake in Terengganu, Pergau in Kelantan, Batang Ai and Bakunin Sarawak, Temenggorin Perak as well as Pedu and Muda in Kedah [8, 9].

There is no comprehensive inventory on all lakes in Malaysia [8, 9]. A preliminary assessment showed that there were over 90 lakes in Malaysia which covered an area of over 100,000 ha and contained more than 30 billion cubic meters of water [8, 9]. Table 4shows the lake inventory based on states in Malaysia [8]. This list of lakes inventory did not, however include ox-bow lakes and ex-mining ponds over the nation. It was estimated that there were approximately more than 4,000 ex-tin mining ponds in the country with area of 16,000 ha [8]. Besides, the inventory excluded new hydroelectric projects like Bakun Dam in Sarawak [8].Bakun Dam alone has 695 km^2 of reservoir surface area with gross storage volume of 44 billion cubic meters [11].

3.1. Pollution and eutrophication of lakes and reservoirs

In Malaysia, water quality of a lake or reservoir can be influenced by external inputs entering the lake or reservoir from the watershed as well as the in-lake ecosystem, nutrients cycling and internal loading. External inputs can be organic and inorganic pollutants as well as nutrients which cause deterioration of water quality from point and non-point sources. Point sources include discharge from domestic and municipal wastewater, agricultural effluents and industrial wastewaters whereas non-point sources includes urban/stormwater runoff, agricultural runoff, septic tank overflow and construction sites runoff. Excessive input of nutrients such as phosphorus and nitrogen into the lakes and reservoirs will lead to eutrophication. Eutrophication is the process by which a water body becomes uncontrollably rich and abundance in aquatic plants such as algae and aquatic macrophytes (water weeds) [12] due to plant nutrients enrichment, especially phosphorus and nitrogen as dissolved solutes and as compounds bound to organic and inorganic matters [13] from natural and anthropogenic sources [12]. Eutrophication will induce an undesirable disturbance to the balance of organisms present in the water and to the quality of water.

According to adesk study on the Status of Eutrophication of Lakes in Malaysia by National Hydraulic Research Institute of Malaysia, NAHRIM [14], 56 lakes or 62 percent of the 90 lakes and reservoirs evaluated were eutrophic while 34 lakes or 38 percent were mesotophic. These lakes were assessed based on the Trophic State Index (TSI) values, computed adopting the landuse and TP relationships. The lakes and reservoirs were graded as good, medium or bad based on allowable nutrient loadings which were corresponded to Carlson's TSI values. Lakes graded as good had TSI values of less than 37.4 whereas TSI values exceeded 47.4 were graded as bad. TSI is not the same as a water quality index. The terms 'good', 'medium' or 'bad' refers to the state of lake in respect to its biological productivity and not 'good water quality', slightly polluted water' or 'polluted water' as detailed in the water quality index. However, if the lakes are confirmed to be eutrophic, steps are needed to tackle the deterioration of water quality. Table 5shows all 90 lakes and reservoirs inventoried in Malaysia with their TSI values and lake status [14-15].

Name/ State/ Federal Territory	River System	Purpose	Area (km²)	Storage capacity (Mm³)	TSI			Lake Status
					LU	Chl a	TP	
State: Perak								
Air Kuning (Perak)	Sg. Ranting/ Sg. Perak	W	n.a.	1.8	60.30	n.a.	n.a.	B
Bersia	Sg. Perak	H, F	5.7	70	60.30	n.a.	n.a.	B
Bukit Merah	Sg. Kurau	I, W	41	75	60.30	n.a.	n.a.	B
Chenderoh	Sg. Perak	H, F	25	95.4	75.44	n.a.	n.a.	B
Gopeng	Sg. Gopeng/ Sg. Perak	SR	n.a.	n.a.	50.00	n.a.	n.a.	B
Jor	Batang Padang/ Sg. Perak	H	0.5	3.9	60.30	n.a.	n.a.	B
Kenering	Sg. Perak	H, F	60	352	60.30	n.a.	n.a.	B
Mahang	Sg. Mahang/ Sg. Perak	H	0.1	0.4	50.00	n.a.	n.a.	B
Temenggor	Sg. Perak	H, F	152	6168	45.85	n.a.	n.a.	M
Kinta	Sg. Kinta/ Sg. Perak	Exm	n.a.	n.a.	68.24	n.a.	n.a.	B
TasikRaban	Sg. Perak	Re	0.375	n.a.	75.44	n.a.	n.a.	B
State: Selangor								
Air Kuning (Selangor)	Sg. Air Kuning/ Sg. Damansara	Re	0.04	0.06	45.85	n.a.	n.a.	M
Batu	Sg. Kelang	F, W	2.5	36	45.85	n.a.	n.a.	M
Klang Gates	Sg. Kelang	W, F	2.25	28.51	45.85	n.a.	n.a.	M
Langat	Sg. Langat	W	1.75	35.49	45.85	n.a.	n.a.	M
Meru	Sg. Subang	W	1	3.5	50.00	n.a.	n.a.	B
Semenyih	Sg. Langat	W	2.5	62.6	45.85	n.a.	n.a.	M
Sungai Baru	Sg. Baru/ Sg. Klang	Re	0.05	0.15	45.85	n.a.	n.a.	M
Sungai Tinggi	Sg. Tinggi/ Sg. Selangor	W	n.a.	107.5	45.85	n.a.	n.a.	M
Sg. Selangor	Sg. Selangor	W	n.a.	235	45.85	n.a.	n.a.	M
Tasik The Mines	Sg. Kuyoh/ Sg. Kelang	Exm, Re	n.a.	n.a.	60.30	n.a.	n.a.	B
TasikTitiwangsa	Sg. Bunus/ Sg. Kelang	Re	0.125	n.a.	-	42.43	n.a.	M

Name/ State/ Federal Territory	River System	Purpose	Area (km^2)	Storage capacity (Mm^3)	TSI			Lake Status
					LU	Chl a	TP	
TasikKundang	Sg. Kundang/ Sg. Selangor	Re	n.a.	n.a.	60.30	n.a.	n.a.	B
TasikAman	Sg. Kelang	Re	0.0224	n.a.	-	91.82	n.a.	B
Damansara	Sg. Damansara	W	0.04	0.009	64.15	n.a.	n.a.	B
Sg. Batu	Sg. Batu	F, W	1.1	2.5	69.39	n.a.	n.a.	B
State: Pahang								
Anak Endau	Sg. Anak Endau	I, W	7.2	38	45.85	n.a.	n.a.	M
Pontian	Sg. Pontian	I, W	20	40	60.30	n.a.	n.a.	B
RepasBaru	Sg. Rengas/ Sg. Pahang	SR	0.05	0.4	68.24	n.a.	n.a.	B
Repas Lama	Sg. Bentong/ Sg. Pahang	SR	n.a.	n.a.	68.24	n.a.	n.a.	B
Sultan Abu Bakar	Sg. Sempam/ Sg. Perak	H	0.5	6.7	68.24	n.a.	n.a.	B
Chereh Dam	Sg. Chereh/ Sg. Kuantan	W	54	250	45.85	n.a.	n.a.	M
TasikChini	Sg. Pahang	N	2	2	68.24	n.a.	n.a.	B
TasikBera	Sg. Bera/ Sg. Pahang	N	6	0.61	68.24	n.a.	n.a.	B
UluLepar	Sg. Lepar/ Sg. Pahang	N	4.69	18	60.30	n.a.	n.a.	B
Bintau	Sg. Kertau/ Sg. Pahang	N	0.25	n.a.	68.24	n.a.	n.a.	B
State: Kelantan								
Bukit Kuang	Sg. Kuang/ Sg. Kelantan	I, W	4.04	14.3	68.24	n.a.	n.a.	B
Pergau (Kuala Yong)	Sg. Pergau	H, F	4.3	62.5	45.85	n.a.	n.a.	M
RantauPanjang	Sg. Golok	W	3	n.a.	75.44	n.a.	n.a.	B
State: Johor								
Bekok	Sg. BatuPahant	F, W	8.75	32	75.44	n.a.	n.a.	B
Congkok	Sg. Tenglu	W	0.5	0.954	45.85	n.a.	n.a.	M
GunongLedang	Sg. Muar	W	0.75	0.3	45.85	n.a.	n.a.	M

http://dx.doi.org/10.5772/58969

Name/ State/ Federal Territory	River System	Purpose	Area (km^2)	Storage capacity (Mm^3)	TSI			Lake Status
					LU	Chl a	TP	
Juaseh	Sg. Juaseh/ Sg. Muar	W	n.a.	33.2	55.03	n.a.	n.a.	B
Labong	Sg. Endau	I, W	4.25	11.54	45.85	n.a.	n.a.	M
Layang (Lower)	Sg. Layang/ Sg. Johor	W	n.a.	11.63	68.24	n.a.	n.a.	B
Layang (Upper)	Sg. Layang/ Sg. Johor	W	n.a.	45	-	n.a.	100.00	B
Lebam	Sg. Lebam	W	n.a.	3.1	-	n.a.	95.95	B
Linggiu	Sg. Linggiu/ Sg. Johor	W	50	772	-	n.a.	57.73	B
Machap	Sg. Benut	F, W	9.09	12.3	75.44	n.a.	n.a.	B
Sembrong	Sg. BatuPahat	F, W	8.5	18	68.24	n.a.	n.a.	B
Pontian Kecil	Sg. Pulai	W	1.75	n.a.	45.85	n.a.	n.a.	M
PulaiBesar	Sg. Pulai	W	0.625	n.a.	45.85	n.a.	n.a.	M
State: Kedah								
Ahning	Sg. Ahning/ Sg. PdgTerap	W, I	n.a.	275	50.00	n.a.	n.a.	B
Malut	Sg. Malut	W	0.5	7.16	50.00	n.a.	n.a.	B
Muda	Sg. Muda	I	26	120	45.85	n.a.	n.a.	M
Padang Saga	Sg. Ulu Melaka	I, W	0.05	0.2	50.00	n.a.	n.a.	B
Pedu	Sg. Kedah	I	65	860	55.03	n.a.	n.a.	B
Dayang Bunting	Sg. Dayang Bunting	N	0.375	n.a.	45.85	n.a.	n.a.	M
Beris	Sg. Beris/ Sg. Muda	W	13.7	1224	50.00	n.a.	n.a.	B
Federal Territory: Labuan								
Bukit Kuda	Sg. Bangat	W	n.a.	4.78	60.30	n.a.	n.a.	B
Kerupang	Sg. Kerupang	W	n.a.	0.21	60.30	n.a.	n.a.	B
Pagar	Sg. Pagar	W	n.a.	0.41	60.30	n.a.	n.a.	B
State: Melaka								
Air Keruh	Sg. Melaka	Re	0.5	n.a.	60.30	n.a.	n.a.	B
Asahan	Sg. Kesang	W	0.75	0.7	50.00	n.a.	n.a.	B
Durian Tunggal	Sg. Melaka	W	3.5	32.6	75.44	n.a.	n.a.	B

Name/ State/ Federal Territory	River System	Purpose	Area (km^2)	Storage capacity (Mm^3)	TSI			Lake Status
					LU	Chl a	TP	
Jus	Sg. Batang Melaka	W	4	48	68.24	n.a.	n.a.	B
State: Negeri Sembilan								
Kelinci	Sg. Kelinchi/ Sg. Pahang	W	n.a.	50	50.00	n.a.	n.a.	B
Pedas	Sg. Beringin/ Sg. Linggi	W	n.a.	0.525	50.00	n.a.	n.a.	B
Sungai Terip	Sg. Terip/ Sg. Linggi	W, I	2.25	48	-	n.a.	62.9	B
Upper Muar	Sg. Muar/ Sg. Pahang	W	n.a.	53	50.00	n.a.	n.a.	B
Gemencheh	Sg. Gemencheh/ Sg. Muar	W	n.a.	30.8	55.03	n.a.	n.a.	B
State: Pulau Pinang								
Air Hitam	Sg. Air Hitam/ Sg. Pinang	W	0.25	2.6	45.85	n.a.	n.a.	M
Mengkuang	Sg. Mengkuang/ Kulim/ Perai	W	0.625	23.6	75.44	n.a.	n.a.	B
TelukBahang	Sg. TelukBahang	W	n.a.	21	45.85	n.a.	n.a.	M
Bukit Pancur	Sg. Kecil/ Sg. Kerian	N	0.061	n.a.	45.85	n.a.	n.a.	M
State: Perlis								
TimahTasoh	Sg. Perlis	I, W, F	13.33	40	-	n.a.	56.6	B
TasikMelati	Sg. Perlis	Re	n.a.	n.a.	50.00	n.a.	n.a.	B
State: Sabah								
Babagon	Sg. Babagon	W	n.a.	21.5	45.85	n.a.	n.a.	M
Pinangsoo	n.a.	W	n.a.	0.24	45.85	n.a.	n.a.	M
Sepagaya	Sg. Silibukan	W	n.a.	2.5	45.85	n.a.	n.a.	M
Tenom	Sg. Pedas	H	n.a.	4.7	45.85	n.a.	n.a.	M
Timbangan	Sg. Kalumpang	W	n.a.	0.67	45.85	n.a.	n.a.	M
Ox-bow	Sg. Kinabatangan	N	n.a.	n.a.	45.85	n.a.	n.a.	M
State: Sarawak								
Batang Ai	Sg. Batang Ai	H	n.a.	2800	45.85	n.a.	n.a.	M
Sika (Bintulu)	Sg. Sika	W	n.a.	3280	45.85	n.a.	n.a.	M

Name/ State/ Federal Territory	River System	Purpose	Area (km^2)	Storage capacity (Mm^3)	TSI			Lake Status
					LU	Chl a	TP	
LoaganBunut	Sg. Tera	N	n.a.	n.a.	45.85	n.a.	n.a.	M
TasikBiru	Sg. Sarawak	Exm	n.a.	n.a.	45.85	n.a.	n.a.	M
State: Terengganu								
Kenyir	Sg. Terengganu	H, F	369	13600	-	-	73.5	B
Puteri/ Bukit Besi	Sg. Paka	Exm, Re	1.8	n.a.	60.30	n.a.	n.a.	B
Federal Territory: Putrajaya								
TasikPutrajaya	Sg. Chua/ Sg. Langat	Re	7.5	45	-	27.07	60.6	M

Notes: W – Water supply; I – Irrigation; H – Hydropower; F – Flood Control; Re – Reclamation; SR – Silt Retention; N – Natural; Exm – Ex-Mining Pool; LU – Land Use (%);Chl a – Chlorophyll a; TP – Total Phosphorus; G – Good; M – Medium; B – Bad.

Table 5. Inventory of Lakes and Reservoirs in Malaysia [14-15].

3.2. Lake and reservoir water quality standards

In Malaysia, there is no specific national standard or index for lake water quality as of year 2014. Since lakes are located within a river basin (although lakes have their own lake basins), water quality standards and classification used are of surface water. Normally, analysis on lake water quality is executed based on the DOE's WQI classification as shown in Table 1 [1], DOE Water Quality Classification based on Water Quality Index as shown in Table 3 [1] and National Water Quality Standards for Malaysia [1]. Although there is no standards and index for lakes in Malaysia, Putrajaya has developed its own standards as stated in Table 6 [17].

Parameter	Unit	Standards
pH		6.5 – 9.0
Temperature	°C	30°C, Normal ±
Colour	TUC	150
Conductivity	μS/cm	1000
Hardness	mg/l	250
Turbidity	NTU	50
Transparency (Secchi)	m	0.6
Taste	-	No Objectionable Taste
Odour	-	No Objectionable Odour

Parameter	Unit	Standards
Floatables	-	No Visible Floatables
Chlorphyll-a	μg/l	0.7
Dissolved oxygen (DO)	mg/l	5 – 7
Biochemical oxygen demand (BOD)	mg/l	3
Chemical oxygen demand (COD)	mg/l	25
Total suspended solid (TSS)	mg/l	150
Ammoniacal nitrogen (AN)	mg/l	0.3
Total Phosphorus (TP)	mg/l	0.05
Nitrate (NO_3-N)	mg/l	7
Nitrite (NO_2-N)	mg/l	0.04
Oil and grease (O&G)	mg/l	1.5
Aluminum	mg/l	<0.05 if pH <6.5 ; <0.1 if pH>6.5
Ammonia	mg/l	0.02 – 0.03
Arsenic	mg/l	0.05
Antimony	mg/l	0.03
Barium	mg/l	1
Beryllium	mg/l	0.004
Boron	mg/l	1
Cadmium	mg/l	0.002
Free chlorine	mg/l	1.5
Chromium, Total	mg/l	0.05
Copper	mg/l	0.02
Cyanide	mg/l	0.02
Iron	mg/l	1
Lead	mg/l	0.05
Manganese	mg/l	0.1
Mercury	mg/l	0.0001
Nickel	mg/l	0.02
Sulphate	mg/l	250
Zinc	mg/l	5

Parameter	Unit	Standards
Microbiological Constituents		
Faecal coliform	counts/100ml	100
Total coliform	counts/100ml	5000
Salmonella	counts/l	0
Enteroviruses	PFU/l	0
Radioactivity		
Gross-alpha	Bq/l	0.1
Gross-Beta	Bq/l	1
Radium-226	Bq/l	<0.1
Strontium-90	Bq/l	<1
Organics		
Carbon Chloroform Extract	μg/l	500
MBAS/ BAS	μg/l	500
Oil & Grease (Mineral)	μg/l	-
Oil & Grease (Emulsified Edible)	μg/l	-
PCB	μg/l	0.1
Phenol	μg/l	10
Aldrin/ Dieldrin	μg/l	0.02
BHC	μg/l	2
Chlordane	μg/l	0.08
t-DDT	μg/l	0.1
Endosulfan	μg/l	10
Heptachlor/ Epoxide	μg/l	0.05
Lindane	μg/l	2
2,4-D	μg/l	70
2,4,5-T	μg/l	10
2,4,5-TP	μg/l	4
Paraquat	μg/l	10

Table 6. The Putrajaya Ambient Lake Water Quality Standards [17].

Besides WQI, lakes and reservoirs in Malaysia can be classified based on their trophic state calculated using Carlson's TSI involving parameters of chlorophyll-a, phosphorus or Secchi Disk depth. TSI classification of lakes and reservoirs is as shown in Table 7 [18].

TSI	Chla (ug/l)	SD (m)	TP (ug/l)	Attributes
<30	<0.95	>8	<6	Oligotrophy: Clear water, oxygen throughout the year in the hypolimnion
30-40	0.95-2.6	8-4	6-12	Hypolimnia of shallower lakes may become anoxic
40-50	2.6-7.3	4-2	12-24	Mesotrophy: Water moderately clear; increasing probability of hypolimnetic anoxia
50-60	7.3-20	2-1	24-48	Eutrophy: Anoxic hypolimnia, macrophyte problems possible
60-70	20-56	0.5-1	48-96	Blue-green algae dominate, algal scums and macrophyte problems
70-80	56-155	0.25-0.5	96-192	Hypereutrophy: (light limited productivity). Dense algae and macrophytes
>80	>155	<0.25	192-384	Algal scums, few macrophytes

Table 7. Trophic State Index (TSI) classification [18].

3.3. Lakes and reservoirs management

The current management of lakes and reservoirs in Malaysia are fragmented with different management agencies or lake managers, each having different and often competing, sectoral agendas [9]. There is no definite policy on lake management and it is neither binding nor mandatory [19]. As such, series of initiatives have been taken towards sustainable management of lakes and reservoirs in Malaysia. In 2009, Academy of Sciences Malaysia (ASM) together with NAHRIM had embarked on the development of Strategies for the Sustainable Development and Management of Lakes and Reservoirs in Malaysia with emphasis on Integrated Lake Basin Management (ILBM) which was first articulated by the International Lake Environment Committee (ILEC). The Strategy Plan for the Sustainable Management of Lakes and Reservoirs in Malaysia was presented and was endorsed during the 7th National Water Resources Council (NWRC) meeting on 1 November 2012[8, 9].

Besides the development of Strategy Plan, initiative had been taken on the development of Lake Briefs based on the format by ILEC. The main objective of Lake Brief is to assist lake managers and stake holders in preparation for gathering information and data relating to lake sand reservoirs towards the development of effective lake basin management plan based on the context of ILBM. Three series of Lake Briefs have been developed in 2010, 2011 and 2012 involving a total of 26 lakes and reservoirs/dam [9, 20-21]. First series of Lake Brief entitled Managing Lakes and their Basins for Sustainable Use in Malaysia, published in 2010 by ASM with support from NRE and NAHRIM involving eight lakes whereas second and third series of Lake Briefs documented by NAHRIM in 2011 and 2012, involved eight and ten lakes and reservoirs, respectively. Lakes and reservoirs documented in the respective Lake Brief are as exhibited in Table 8 [9, 20-21].

Lake Brief Series I (2010)	Lake Brief Series II (2011)	Lake Brief Series III (2012)
• Bukit Merah, Perak	• TasikBera, Pahang	• Batang Ai, Sarawak
• TasikKenyir, Terengganu	• Paya Indah Wetlands (PIW), Selangor	• Empangan Bukit Kwong, Kelantan
• LoaganBunut, Sarawak	• EmpanganBeris, Sik, Kedah	• Durian Tunggal, Melaka
• TasikPedudanMuda, Kedah	• EmpanganSemberong, BatuPahat, Johor	• Tasik Taiping, Perak
• TasikPutrajayadan Wetlands	• Tasik Ringlet, Cameron Highlands, Pahang	• Tasik FRIM dan Wetlands, Selangor
• TasikChini, Pahang	• EmpanganChenderoh, Perak	• TasikJor, Perak
• TasikTerip, Negeri Sembilan	• EmpanganKlang Gate	• TasikPergau, Kelantan
• TasikTimahTasoh, Perlis	• EmpanganSg. Selangor, Selangor	• Babagon, Sabah
		• TasikSubang, Selangor
		• EmpanganLangat, Selangor

Table 8. List of lakes and reservoirs in Lake Brief reports [9, 20-21].

A systematic approach of lake management plan involving all stakeholders especially community participation to cultivate the sense of ownership should be established and implemented. The successful case of lake management is the Putrajaya Lake and Wetland. Putrajaya has a comprehensive Catchment Development and Management Plan for Putrajaya Lake (CDMPPL) which incorporates various management guidelines and studies, aim to achieve and maintain the high water quality objective set for Putrajaya Lake [22]. The Putrajaya Drainage Master Plan 1996 was developed to preserve the urban ecological values within the wetlands and lake areas and to mitigate the runoff and pollutants exports to external catchments whereas the Putrajaya Integrated Stormwater Management Guidelines set strategies that include stormwater drainage, managing stormwater as a resource, protection of receiving water quality and protection of downstream ecological health [23].

4. Coastal, estuarine and island marine water

Malaysia is one of richest in marine biodiversity in the world where these marine resources are essential contributions to the livelihood and sustenance of its people. Major marine ecosystems in Malaysia include coral reefs, seagrasses, mangroves, mudflats and estuaries [24]. However, these coastal, estuarine and island marine waters are increasingly impacted by pollutants and materials discharged from land-based activities due to urbanization, development and increase in population, resulted in the deterioration of water quality. These threats are generally from point or non-point sources. Point sources of pollutant include sewage and municipal wastewater as well as industrial wastewaters [24]. Non-point sources are runoffs from urban, agriculture, land clearing and construction activities as well as deposition from atmospheric sources. Besides land-based pollution, the marine environment is exposed to threats from the shipping activities, offshore oil and gas exploration and exploitation activities [24].

In Malaysia, the marine water quality is monitored by the Department of Environment (DOE) in Peninsular Malaysia since 1978 and 1985 in Sabah and Sarawak. The aim is to identify the marine water quality status and to determine the degree of pollution from both the land-based sources as well as the sea-based sources that can pose threats to the marine resources which potentially contribute to the stability and diversity of the marine ecosystem [16, 25]. In 2012, there were around 168 coastal and 78 estuary monitoring stations and 76 islands being monitored with a total of 579 samples from coastal, 325 samples from estuary and 190 samples collected from island monitoring stations [3]. Analyses on data were based on the newly developed Marine Water Quality Index (MWQI). As for coastal water quality, only 155 stations of 168 stations were analysed for MWQI [18]. Figure 2illustrates the trend in terms of MWQI for 2010-2012 [3]. It was observed that three stations (1.9 percent) were categorised as excellent, 32 stations (20.6percent) as good, 111 stations (71.6percent) as moderate and nine stations (5.8percent) as poor in year 2012. The number of excellent and good marine water quality stations had decreased, while the stations with moderate water quality had increased [3].

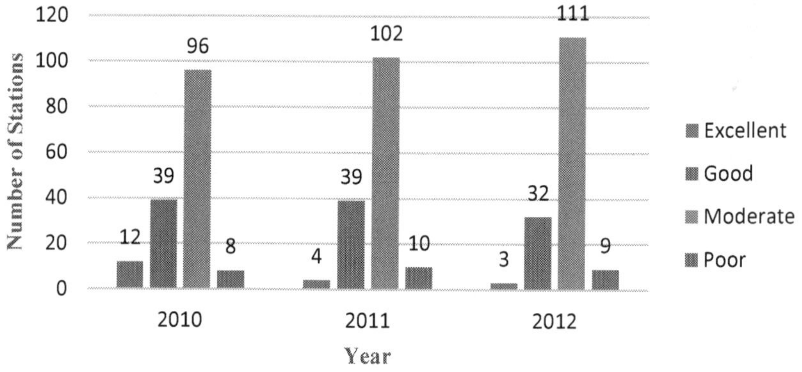

Figure 2. Marine Water Quality Index Trend for Coastal Areas in Malaysia from 2010 to 2012 [3].

Of the 78 estuary stations monitored in 2012, only 69 were analysed for MWQI. The comparison of MWQI trend from 2010 to 2012 in Malaysia is as depicted in Figure 3 [3]. It was observed that there was a slight improvement for the good stations as the good stations had increased from 8.7percent in 2011 to 11.6percent in 2012. Besides, moderate stations showed some decrements from 50 stations in 2011 to 48 stations in 2012, while excellent and poor stations remained the same for 2011-2012 [3].

As for the 76 islands monitored in 2012, 3 islands were classified as development islands, 32 island as resort islands, 29 stations as marine park islands whereas 12 stations were protected

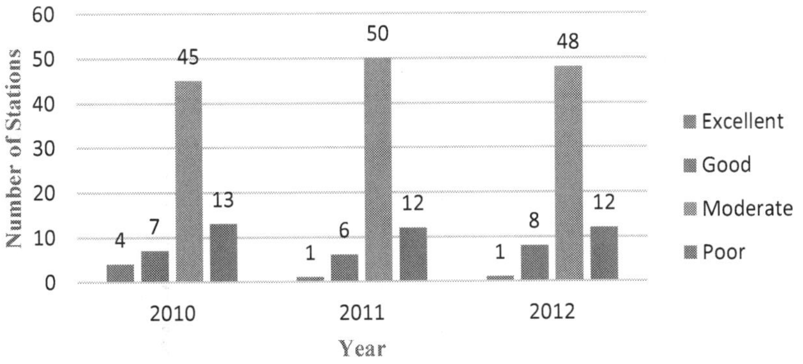

Figure 3. Marine Water Quality Index Trend for Estuary Areas in Malaysia from 2010 to 2012 [3].

islands [3]. Development islands are Labuan Island, Pulau Pinang and Langkawi, resort islands are like Sipadan in Sabah, Dayang Bunting in Kedah and Pangkor in Perak while marine park islands are islands such as PulauPayar in Kedah, Tioman Island in Pahang, Perhentian in Terengganu and Sapi in Sabah. Protected islands are like Pulau Talang-talang Besar in Sarawak, Pulau Arang in Negeri Sembilan, Kukup in Johor and Pulau Panjang in Kelantan [3]. From 76 islands with 93 island stations monitored, 86 were analysed for MWQI [3]. Figure 4 shows the comparison of MWQI for islands in Malaysia between year 2011 and 2012 [3]. It was observed that there is a slight decrement of station with good water quality from 17 stations in 2011 to 13 stations in 2012. 18 stations (20.9percent) in 2012 were categorised as good, while 52 stations (60.5percent) as moderate and were almost the same as recorded in 2011. Stations with poor water quality were very least with only 3 stations in 2012 and 2 stations in 2011 [3].

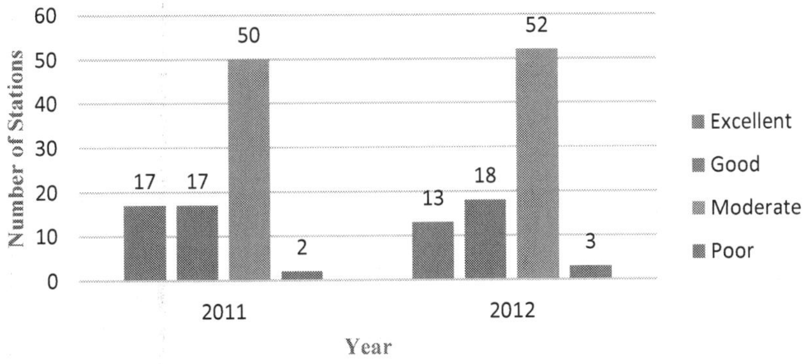

Figure 4. Marine Water Quality Index Trend for Islands (2011 – 2012) [3].

4.1. Marine water quality standards

Water quality at the coastal areas, estuarine and islands are analysed based on the newly developed MWQI by the DOE. The Index is developed based on an opinion poll approach using the Delphi technique involving experts from various group of stakeholders including professionals, practitioners, academicians related to marine water quality management, specialists from marine resource management, marine ecosystems, legislations and regulations, consultants, marine biologist, hydrologist as well as experts from governmental and non-governmental organizations [24]. The MWQI is used as a way to reflect the marine water quality status and category. The index was developed based on seven main parameters consist of dissolved oxygen, DO (0.2), ammonia, NH_3 (0.16), faecal coliform (0.14), total suspended solids, TSS (0.14), Oil and Grease, O&G (0.13), nitrate, NO3 (0.12) and phosphate, PO4 (0.11) with corresponding weightages in parenthesis, reflecting contributions of the respective parameters to the overall water quality [24]. The resulting MWQI is classified into four categories of water quality consisting excellent with index value of 90 – 100, good (80 - <90), moderate (50 - <80) and poor (0 - <50) [24].

Apart from MWQI, Malaysia Marine Water Quality Criteria and Standard (MWQCS) is also developed with four different classes (Class 1, Class 2, Class 3 and Class E) for four different beneficial uses as shown in Table 9 [16, 24]. Class 1 and Class 2 in the MWQCS require MWQI values equal to or greater than 80 and are categorized at least as 'good' water quality [16, 24]. The Class 3 waters of the MWQCS are generally within the moderate category of the MWQI [16, 26]. The application of MWQI to Class E waters requires the samples taken must correspond to a salinity of not less than 25 ppt [24].

Parameter	Class 1	Class 2	Class 3	Class E
Beneficial Uses	Preservation, Marine Protected Areas, Marine Parks	Marine Life, Fisheries, Coral Reefs, Recreational and Mariculture	Ports, Oil & Gas Fields	Mangroves Estuarine & River-mouth Water
Temperature	≤ 2°C increase over maximum ambient	≤ 2°C increase over maximum ambient	≤ 2°C increase over maximum ambient	≤ 2°C increase over maximum ambient
Dissolved oxygen (mg/L)	>80% saturation	5	3	4
Total suspended solid (mg/L)	25 mg/L or ≤ 10% increase in seasonal average, whichever is lower	50mg/L (25 mg/L) or ≤ 10% increase in seasonal average, whichever is lower	100 mg/L or ≤ 10% increase in seasonal average, whichever is lower	100 mg/L or ≤ 30 % increase in seasonal average, whichever is lower
Oil and grease (mg/L)	0.01	0.14	5	0.14
Mercury (μg/L)	0.04	0.16 (0.04)	50	0.5
Cadmium (μg/L)	0.5	2 (3)	10	2
Chromium (VI) (μg/L)	5	10	48	10

Parameter	Class 1	Class 2	Class 3	Class E
Copper (μg/L)	1.3	2.9	10	2.9
Arsenic (III) (μg/L)	3	20 (3)	50	20 (3)
Lead (μg/L)	4.4	8.5	50	8.5
Zinc (μg/L)	15	50	100	50
Cyanide (μg/L)	2	7	20	7
Ammonia (unionized) (μg/L)	35	70	320	70
Nitrite (NO2) (μg/L)	10	55	1,000	55
Nitrate (NO3) (μg/L)	10	60	1,000	60
Phosphate (μg/L)	5	75	670	75
Phenol (μg/L)	1	10	100	10
Tributyltin (TBT) (μg/L)	0.001	0.01	0.05	0.01
Faecal coliform (Human health protection for seafood consumption) – Most Probable Number (MPN)	70 faecal coliform 100mL-1	100 faecal coliform 100mL-1 & (70 faecal coliform 100mL-1)	200 faecal coliform 100mL-1	100 faecal coliform 100mL-1 & (70 faecal coliform 100mL-1)
Polycyclic Aromatic Hydrocarbon (PAHs) (ng/g)	100	200	1000	1000
*IMWQS in parentheses are for coastal and marine water areas where seafood for human consumption is applicable				

Table 9. Malaysia Marine Water Quality Criteria and Standards [16, 24].

5. Groundwater

There are the freshwater in rivers, lakes and aquifers, and groundwater resources account for about 99 percent[26]. Despite the abundance of groundwater, it only accounts for 3 percent of total water use for domestic, agricultural, and industrial sectors. Lack of information to indicate groundwater source, perception that the supply is non-sustainable and lack of local expertise on groundwater are factors of under utilization of groundwater resources. Furthermore, leachates from the disposal of waste in landfills pose a threat to human health and environment [27].

On top of that, costly installation of groundwater extraction system, plus readily water supply available to over 90 percent of the communities, groundwater development is said to be ignored. So far groundwater is only extracted through well for domestic use and irrigation in very rural areas. This situation continues until water crisis occurred in 1997. After the water crisis, DOE has taken preliminary steps to explore and determine the quality and distributions of groundwater through the national groundwater-monitoring programme.

5.1. Groundwater wells and quality

By 2012, 107 quality monitoring wells had been established, as shown in Table10[1]. Selection of the sites were done according to the surrounding land uses which were agricultural, urban or suburban, rural, industrial, solid waste landfills, golf courses, radioactive landfill, animal burial areas, municipal water supply and examining areas (gold mine).

Category	Number of Wells
Agricultural Areas	12
Urban/Suburban Areas	11
Industrial Sites	18
Solid Waste Landfills	24
Golf Courses	7
Rural Areas	3
Ex-mining Areas (Gold Mine)	3
Municipal Water Supply	7
Animal Burial Areas	14
Aquaculture Farms	6
Radioactive Landfill	1
Resorts	1
TOTAL	107

Table 10. Number of groundwater wells by land use category [1]

According to [1], 361 water samples were taken from these monitoring wells for the analysis of physical, chemical and biological characteristics. These include total dissolved solids (TDS), pH, temperature, dissolved oxygen and conductivity for physical characteristic; volatile organic compounds (VOCs), pesticides, heavy metals, anions and phenolic compounds for chemical characteristic; and bacteria (coliform) for biological characteristic.The assessment of groundwater quality was based on the percentage of samples exceeded the National Guidelines for Raw Drinking Water Quality 2000 benchmark. In 2012, the results of monitoring showed that all stations were within the benchmark except for arsenics (As), iron (Fe), manganese (Mn), total coliform and phenol. Total coliform was categorized as exceeded the benchmark and in fact were recorded at all stations, followed by phenol, Fe, Mn and As [1].

5.2. Groundwater quality standard

The Ground Water Quality Standards for Malaysia is still not established, but considering potential of groundwater as an alternative source for surface water, DOE had started the National Groundwater Monitoring Programme to determine the groundwater quality status. The groundwater quality status was determined based on the National Guidelines for Raw DrinkingWater Quality, as shown in Table 11, as the benchmark [1].

Parameter	Symbol	Benchmark
Sulphate	SO4	250 mg/l
Hardness	CaCO3SO	500 mg/l
Nitrate	NO3SO	10 mg/l
Coliform	-	Must not be detected in any 100 ml sample
Manganese	Mn	0.1 mg/l
Chromium	Cr	0.05 mg/l
Zinc	Zn	3 mg/l
Arsenic	As	0.01 mg/l
Selenium	Se	0.01 mg/l
Chloride	Cl	250 mg/l
Phenolics	-	0.002 mg/l
TDS	-	1000 mg/l
Iron	Fe	0.3 mg/l
Copper	Cu	1.0 mg/l
Lead	Pb	0.01 mg/l
Cadmium	Cd	0.003 mg/l
Mercury	Hg	0.001 mg

Table 11. National Guidelines for Raw Drinking Water Quality - revised December 2000 [1]

6. Drinking water

According to [28], Malaysia receives more than 25,000 cubic meters of renewable water per capita annually from its extensive river system that consists of more than 150 rivers. The amount of renewable water that Malaysia receives far exceeds that of many other parts of the world. Thus, over the past 200 years, consumers have the convenience of running water at the turn of a tap.

In Malaysia, the most tapped raw water sources are rivers, which are technically under the jurisdiction of the respective state governments. In the past, conventional treatment systems were employed to treat raw surface water. Typically, this kind of treatment system provides very basic treatment which includes screening, coagulation and flocculation, sand filtration, disinfection (chlorination) and fluoridation. As development becomes more rampant especially industrialization, river water quality becomes seriously deteriorated plus the water characteristic becomes more complex with the existence of new contaminants. At this time, conventional treatment system is not capable to remove these contaminants and as a result they might enter the distribution and supply network. To manage this problem, the Environmental Quality Act, 1974, prescribes more stringent regulatory compliance for wastewater discharging premises located upstream of a water intake point [29]. In spite of this, not all contaminants are covered under the Act, therefore the risk of contamination cannot be totally eradicated.

6.1. Drinking water standard

In 1985, DOE has developed the Drinking Water Quality Standards for Malaysia, as shown in Table 12 below:

Parameter	Group	MALAYSIA STANDARD mg/l (unless otherwise stated)			
		Raw Water		Treated Water	
		Min	Max	Min	Max
Total Coliform	1	0	5000MPN/100 ml	0 in 100 ml	0
E.coli	1	0	5000MPN/100 ml	0 in 100 ml	0
Turbidity	1	0	1000 NTU	0	5 NTU
Color	1	0	300 TCU	0	15 TCU
PH	1	5.50000	9.00000	6.50000	9.00000
Free Residual Chlorine	1	-	-	0.20000	5.00000
Temperature	1	-	-	-	-
Clostridium perfringens (including spores)	1	-	-	0	Absent
Coliform bacteria	1	-	-	-	-
Colony count 22°	1	-	-	-	-
Conductivity	1	-	-	-	-
Enterococci	1	-	-	-	-
Odour	1	-	-	-	-
Taste	1	-	-	-	-

http://dx.doi.org/10.5772/58969

Parameter	Group	MALAYSIA STANDARD mg/l (unless otherwise stated)			
		Raw Water		Treated Water	
		Min	Max	Min	Max
Oxidisability	1	-	-	-	-
TDS	2	0.00000	1500.00000	0.00000	1000.0000
Chloride	2	0.00000	250.00000	0.00000	250.00000
Ammonia	2	0.00000	1.50000	0.00000	1.50000
Nitrat	2	0.00000	10.00000	0.00000	10.00000
Ferum/Iron	2	0.00000	1.00000	0.00000	0.30000
Fluoride	2	0.00000	1.50000	0.40000	0.60000
Hardness	2	0.00000	500.00000	0.00000	500.00000
Aluminium	2	-	-	0.00000	0.20000
Manganese	2	0.00000	0.20000	0.00000	0.10000
COD	2	0.00000	10.00000	-	-
Anionic Detergent MBAS	2	0.00000	1.00000	0.00000	1.00000
BOD	2	0.00000	6.00000	-	-
Nitrite	2	-	-	-	-
Total organic carbon	2	-	-	-	-
Mercury	3	0.00000	0.00100	0.00000	0.00100
Cadmium	3	0.00000	0.00300	0.00000	0.00300
Arsenic	3	0.00000	0.01000	0.00000	0.01000
Cyanide	3	0.00000	0.07000	0.00000	0.07000
Plumbum/Lead	3	0.00000	0.05000	0.00000	0.01000
Chromium	3	0.00000	0.05000	0.00000	0.05000
Cuprum/Copper	3	0.00000	1.00000	0.00000	1.00000
Zinc	3	0.00000	3.00000	0.00000	3.00000
Natrium/Sodium	3	0.00000	200.00000	0.00000	200.00000
Sulphate	3	0.00000	250.00000	0.00000	250.00000
Selenium	3	0.00000	0.01000	0.00000	0.01000
Argentum	3	0.00000	0.05000	0.00000	0.05000
Magnesium	3	0.00000	150.00000	0.00000	150.00000
Mineral Oil	3	0.00000	0.30000	0.00000	0.30000
Chloroform	3	-	-	0.00000	0.20000

Parameter	Group	MALAYSIA STANDARD mg/l (unless otherwise stated)			
		Raw Water		Treated Water	
		Min	Max	Min	Max
Bromoform	3	-	-	0.00000	0.10000
Dibromoklorometana	3	-	-	0.00000	0.10000
Bromodiklorometana	3	-	-	0.00000	0.06000
Fenol/Phenol	3	0.00000	0.00200	0.00000	0.00200
Antimony	3	-	-	0	0.005
Nickel	3	-	-	0	0.02
Dibromoacetonitrile	3	-	-	0	0.1
Dichloroacetic acid	3	-	-	0	0.05
Dichloroacetonitrile	3	-	-	0	0.09
Trichloroacetic acid	3	-	-	0	0.1
Trichloroacetonitrile	3	-	-	0	0.001
Trihalomethanes - Total	3	-	-	-	-
Aldrin / Dealdrin	4	0.00000	0.00003	0.00000	0.00003
DDT	4	0.00000	0.00200	0.00000	0.00200
Heptachlor &Heptachlor Epoxide	4	0.00000	0.00003	0.00000	0.00003
Methoxychlor	4	0.00000	0.02000	0.00000	0.02000
Lindane	4	0.00000	0.00200	0.00000	0.00200
Chlordane	4	0.00000	0.00020	0.00000	0.00020
Endosulfan	4	0.00000	0.03000	0.00000	0.03000
Hexachlorobenzena	4	0.00000	0.00100	0.00000	0.00100
1,2-dichloroethane	4	-	-	0.00000	0.03000
2,4,5-T	4	-	-	0.00000	0.00900
2,4,6-trichlorophenol	4	-	-	0.00000	0.20000
2,4-D	4	0.00000	0.03000	0.00000	0.03000
2,4-DB	4	-	-	0.00000	0.09000
2,4-dichlorophenol	4	-	-	0.00000	0.09000
Acrylamide	4	-	-	0.00000	0.00050
Alachlor	4	-	-	0.00000	0.02000
Aldicarb	4	-	-	0.00000	0.01000
Benzene	4	-	-	0.00000	0.01000

Parameter	Group	MALAYSIA STANDARD mg/l (unless otherwise stated)			
		Raw Water		Treated Water	
		Min	Max	Min	Max
Carbofuran	4	-	-	0.00000	0.00700
MCPA	4	-	-	0.00000	0.00200
Pendimethalin	4	-	-	0.00000	0.02000
Pentachlorophenol	4	-	-	0.00000	0.00900
Permethrin	4	-	-	0.00000	0.02000
Pesticides	4	-	-	-	-
Pesticides - Total	4	-	-	-	-
Polycyclic aromatic hydrocarbons	4	-	-	-	-
Propanil	4	-	-	0.00000	0.02000
Tetrachloroethene and Trichloroethene	4	-	-	-	-
Vinyl chloride	4	-	-	0.00000	0.00500
Gross alpha (α)	5	0.00000	0.1Bq/l	0.00000	0.1Bq/l
Gross beta (β)	5	0.00000	1.0 Bq/l	0.00000	1.0 Bq/l
Tritium	5	-	-	-	-
Total indicative dose	5	-	-	-	-

Table 12. Drinking Water Quality Standard [2]

6.2. Drinking water treatment technology

In recent years, the pollution in rivers and lakes has become increasingly worse, and it has become necessary to take measures toward securing a stable supply of tap water and supplying good-tasting water. Generally, there are two methods used in water treatment, which are the conventional and non-conventional methods. Conventional method is water treatment which undergoes processes such as coagulation, flocculation, sedimentation, and filtration, as shown in Figure 5, while non-conventional method uses more sophisticated equipment.

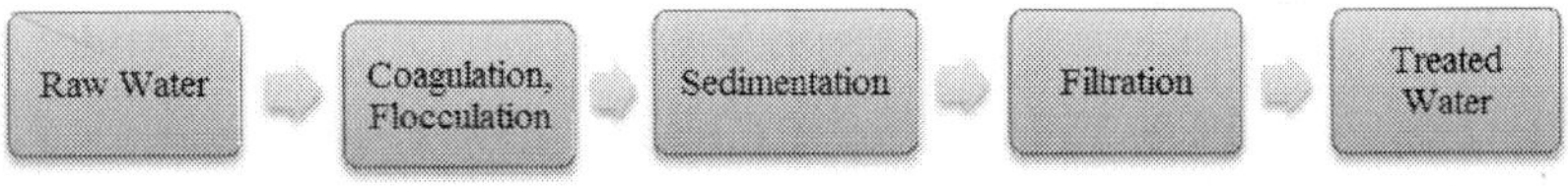

Figure 5. Schematic diagram for conventional treatment plant

In Malaysia, both conventional and non-conventional methods are employed. Generally, if the water is severely contaminated or there is a need for alternative water source, then conventional method is no longer viable for use. At present, Dissolved Air Flotation (DAF) and Actiflo are among the latest conventional technology in use.

In DAF float tank, suspended matter in the feed water is often flocculated with coagulant (such as ferric chloride or aluminum sulfate). These flocculated suspended matters are then floated up to the surface by the adhered bubbles and form a froth layer which is then removed by a skimmer. The froth-free water exits the float tank as the clarified effluent. The tiny bubbles are actually those bubbles that released from the pressure reduction valve when the mixture of clarified effluent and the compressed air are forced to flow through this valve (Figure 6).

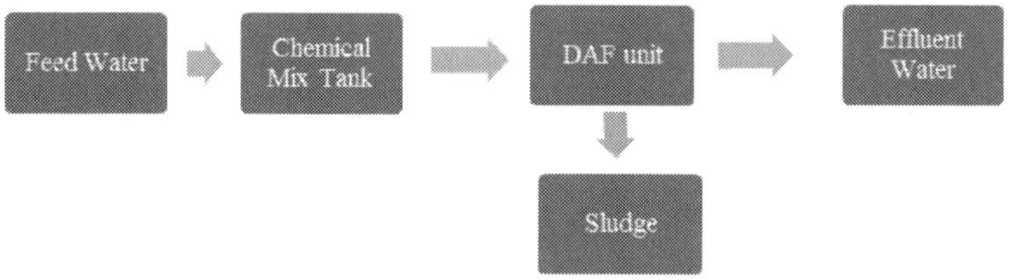

Figure 6. Schematic diagram for DAF technology

One of the non-conventional methods used in Malaysia is membrane technology. Ultrafiltration process in this technology uses membrane modules (Figure 3) as filtration media as compared to sand in conventional filtration plant. At present, one water treatment plant in Bukit Panchor, Pulau Pinang [31] and two plants in Selangor [32] are adopting the application of this membrane technology. Ultrafiltration (UF) is a variety of membrane filtration in which forces like pressure or concentration gradients leads to a separation through a semipermeable membrane. Suspended solids and solutes of high molecular weight are retained in the so-called retentate, while water and low molecular weight solutes pass through the membrane in the permeate. Among the advantages of using the UF technology is, it could produce clean water of high quality, lower operation cost, easily upgradeable system and space reducing compact system.

The progression and development of the water treatment technology are undeniably based on the deteriorating water source quality and current water demand over time resulted from the growth of population and industries. In addition, advancement in water treatment technology minimizes the space usage, increases production yield, more cost effective, and reduced labor.

7. Conclusion

The DOE uses WQI and NWQS to evaluate the status of the river water quality. There is no specific national standard or index for lake water quality analyses for the meantime. Since lakes are located within a river basin (although some say lakes have their own lake basins), the water quality standards and classification for rivers are used. On the other hand, the marine water quality is monitored by the DOE in Peninsular Malaysia since 1978, and 1985 in Sabah and Sarawak, with the aim to identify the marine water quality status and to determine the degree of pollution from both the land-based sources as well as the sea. Water quality at the coastal areas, estuarine and islands are analysed based on the newly developed Marine Water Quality Index (MWQI) by the DOE. After the water crisis in 1997, DOE has taken preliminary steps to explore and determine the quality and distributions of groundwater through the national groundwater-monitoring programme. Consequently, DOE had started the National Groundwater Monitoring Programme to determine the groundwater quality status. For drinking water, DOE has developed the Drinking Water Quality Standards for Malaysia since 1985. Selection of water treatment technology depends on the quality of water source. Among the technologies in use in the country are DAF, Actiflo and membrane technology.

Acknowledgements

The authors would like to express their sincere appreciations to the Department of Environment (DOE) Malaysia for their water quality monitoring work and comprehensive reports to make this book chapter a success. Special thanks are also extended to the other relevant Government's agencies, departments and institutes for their information and materials to enhance the contents of this document.

Author details

Yuk Feng Huang[1], Shin Ying Ang[2], Khia Min Lee[1] and Teang Shui Lee[3*]

*Address all correspondence to: teangshuilee@gmail.com

1 Faculty of Engineering and Science, Universiti Tunku Abdul Rahman, Kuala Lumpur, Malaysia

2 National Hydraulic Research Institute of Malaysia, Selangor, Malaysia

3 Faculty of Engineering, Universiti Putra Malaysia, Selangor, Malaysia

References

[1] Department of Environment. Malaysia: Environmental Quality Act report, Ministry of Science, Technology and the Environment, Putrajaya, Malaysia, 2012.

[2] Department of Environment Malaysia, "Development of Water Quality Criteria and Standards for Malaysia", 1985.

[3] Thomas R, Meybeck M, Beim A. Chapter 7 – Lakes. In Chapman D (ed.) Water Quality Assessment – A Guide to Use of Biota, Sediments and Water Environmental Monitoring – Second Edition.UNESCO, WHO, UNEP, 1996.

[4] Ramadas K. Report on Governance Sector: Position Paper on Governance of Lakes in Malaysia (2008). Strategy Plan: Strategies for the Sustainable Development and Management of Lakes and Reservoirs in Malaysia Volume 2. Kuala Lumpur: ASM and NAHRIM; 2009.

[5] ILEC. Integrated Lake Basin Management: An Introduction.Kusatsu, Japan: International Lake Environment Committee Foundation; 2007. http://www.ilec.or.jp/en/wp/wp-content/uploads/2013/03/ILBM_Report_E_07oct02.pdf (accessed 15 July 2014).

[6] Nakamura M. Improving Basin Governance Toward Integrated Lake/Reservoir Management, A Global Challenge. In Anton A, Zakaria, S, Yusoff, FM, Low, KS, Lee J, Sharip, Z. (eds.) Lake and Reservoir Management in Malaysia: Status and Issues – Proceedings of the Colloquium on Lakes and Reservoir Management 2 & 3 August 2007.

[7] ILEC. Managing Lakes and their Basins for Sustainable Use: A Report of Lake Basin Managers and Stakeholders. Kusatsu, Japan: International Lake Environment Committee Foundation; 2005. http://www.ilec.or.jp/en/wp/wp-content/uploads/2013/03/LBMI_Main_Report_22February20061.pdf (accessed 9 June 2014).

[8] ASM, NAHRIM. PelanStrategi: Strategi Pembangunan danPengurusan Lestari bagiTasikdanEmpangan Air di Malaysia Jilid 1: LaporanUtama. 2009.

[9] ASM. Lake Briefs Synthesis Report. Managing Lakes and Their Basins for Sustainable Use in Malaysia (Lake Briefs Report Series 1). Kuala Lumpur: AkademiSains Malaysia; 2010.

[10] Hidzrami SA. Bukit Merah, Perak Lake Brief.Lakes and Their Basins for Sustainable Use in Malaysia (Lake Briefs Report Series 1). Kuala Lumpur: AkademiSains Malaysia; 2010.

[11] SPU. Sarawak Integrated Water Resources Management Master Plan Study. State Planning Unit, Sarawak; 2008.http://www.siwrs.com.my/modules/iwrm/page.php?id=8&menu_id=0&sub_id=3 (accessed 28 July 2014).

[12] Salameh E, Harahsheh S. Eutrophication Processes in Arid Climates. In Ansari AA, Gill SS, Lanza GR, Rast W (eds.) Eutrophication: Causes, Consequences and Control. New York: Springer; 2011.

[13] UNEP-IETC. Planning and Management of Lakes and Reservoirs: An Integrated Approach to Eutrophication, A Training Module. Osaka: United Nations Environment

Programme International Environmental Training Centre; 2000. ftp://ftp.icess.ucsb.edu/pub/org/limno/data185/mary/UNEP/UNEP_merged2.pdf(accessed 15 July 2014)

[14] NAHRIM. A Desk Study on the Status of Eutrophication of Lakes in Malaysia (2005). Seri Kembangan: National Hydraulic Research Institute of Malaysia; 2009.

[15] ASM, NAHRIM. Strategy Plan: Strategies for the Sustainable Development and Management of Lakes and Reservoirs in Malaysia Volume 3 Part 1. 2009.

[16] DOE. Malaysia Environmental Quality Report 2012.Putrajaya: Department of Environment Malaysia; 2012.

[17] Abdullah Rani AB. Putrajaya Lake and Wetland: Water Quality Status. National Seminar on Ecosystem Management for Lakes and Wetland of Putrajaya; 2013. http://www.mwa.org.my/events/20130205/6.%20Dr%20Rani%20-Putrajaya-%20seminar%202013v4.pdf (accessed 10 October 2013)

[18] Carlson RE and Simpson J.A Coordinator's Guide to Volunteer Lake Monitoring Methods.North American Lake Management Society.TheSecchi Dip-In website; 1996. http://www.secchidipin.org/tsi.htm (accessed16 February 2014)

[19] Mushrifah I. Sustainable Management of Lakes in Malaysian Scenario – A Survey of its Management. Proceedings National Seminar on Research and Management of Lakes and their Basins for Sustainable Use: The Current Status of Selected Lakes in Malaysia 3rd Series.

[20] NAHRIM. PengurusanTasikdanLembangannyasecaraBersepadu di Malaysia: Penyediaan Lake Brief Siri 2 (2010/2011) 10 -11 Mei 2011, Cameron Highlands, Pahang. Seri Kembangan: National Hydraulic Research Institute of Malaysia; 2012.

[21] NAHRIM. Proceedings National Seminar on Research and Management of Lakes and their Basins for Sustainable Use: The Current Status of Selected Lakes in Malaysia 3rd Series 27 & 28 September 2012, Paya Indah Wetlands. Seri Kembangan: National Hydraulic Research Institute of Malaysia; 2013.

[22] Selamat Z. Putrajaya Lake: Sustainable Management of Created Wetland. In Anton A, Zakaria, S, Yusoff, FM, Low, KS, Lee J, Sharip, Z. (eds.) Lake and Reservoir Management in Malaysia: Status and Issues – Proceedings of the Colloquium on Lakes and Reservoir Management.

[23] Majizat A. The Case of Putrajaya Lake and Wetland: Lessons on Preparing and Implementing a Lake and Wetlands Management Plan within an Urbabn Development Framework. Integrated Lake Basin Training Material. Japan: International Lake Environment Committee; 2009. http://wldb.ilec.or.jp/ILBMTrainingMaterials/resources/Putrajaya.pdf (accessed 16 September 2012).

[24] DOE. Managing Our Seas: The Malaysian Marine Water Quality Index. Putrajaya: Department of Environment Malaysia; 2013.

[25] DOE. Malaysia Environmental Quality Report 2010.Putrajaya: Department of Environment Malaysia; 2010.

[26] Government of Malaysia/JICA. National Water Resources Study, Malaysia, Sectoral Report, 1982; Volume 18.

[27] Suratman S., Tawnie I., Sefei, A. Impact of landfills on groundwater in Selangor, Malaysia. ASM Sci. J., 2011 5(2), 101-107.

[28] Keizul A. Hydrology For Sustainable Water Resources Planning, Development and Management in Malaysia. International Conference on Hydrology and Water Resources of Humid Tropics, 24-26 Nov. 1998, Ipoh, Malaysia.

[29] Baginda A. R. A., Zainudin Z. Keynote Paper : Moving Towards Integrated River Basin Management (IRBM) in Malaysia. Institution of Engineers Malaysia (IEM).Proceedings, 11th Annual IEM Water Resources Colloquium 2009, ISBN 978-967-5048-46-3.

[30] Sharaai A. H., Mahmood N. Z., Sulaiman A. H. (2010). Life Cycle Impact Assessment (LCIA) of Potable Water Production in Malaysia: A Comparison among Different Technology Used in Water Treatment Plant. EnvironmentAsia, 3(1), pg 95-102

[31] PBA. Bukit Pancur Treatment Plant. Pulau Pinang: PBA.2006.

[32] Ibrahim N. A. (2008) PuncakNiagaperkenalteknologibarutapis air. Malaysia standards handbook on environment management: MS ISO 14000 Series (2nd ed.)Shah Alam, Malaysia: SIRIM

Chapter 4

Validity and Errors in Water Quality Data — A Review

Innocent Rangeti, Bloodless Dzwairo,
Graham J. Barratt and Fredrick A.O. Otieno

Additional information is available at the end of the chapter

http://dx.doi.org/10.5772/59059

1. Introduction

While it is essential for every researcher to obtain data that is highly accurate, complete, representative and comparable, it is known that missing values, outliers and censored values are common characteristics of a water quality data-set. Random and systematic errors at various stages of a monitoring program tend to produce erroneous values, which complicates statistical analysis. For example, the central tendency statistics, particularly the mean and standard deviation, are distorted by a single grossly inaccurate data point. An error, which is initially identified and is later incorporated into a decision making tool, like a water quality index (WQI) or a model, could subsequently lead to costly consequences to humans and the environment.

Checking for erroneous and anomalous data points should be routine, and an initial stage of any data analysis study. However, distinguishing between a data-point and an error requires experience. For example, outliers may actually be results which might require statistical attention before a decision can be made to either discard or retain them. Human judgement, based on knowledge, experience and intuition thus continue to be important in assessing the integrity and validity of a given data-set. It is therefore essential for water resources practitioners to be knowledgeable regarding the identification and treatment of errors and anomalies in water quality data before undertaking an in-depth analysis.

On the other hand, although the advent of computers and various software have made it easy to analyse large amounts of data, lack of basic statistical knowledge could result in the application of an inappropriate technique. This could ultimately lead to wrong conclusions that are costly to humans and the environment [1]. Such necessitate the need for some basic understanding of data characteristics and statistics methods that are commonly applied in the water quality sector. This chapter, discusses common anomalies and errors in water quality data-sets, methods of their identification and treatment. Knowledge reviewed could assist with building appropriate and validated data-sets which might suit the statistical method under consideration for data analysis and/or modelling.

2. Data errors and anomalies

Referring to water quality studies, an error can be defined as a value that does not represent the true concentration of a variable such as turbidity. These may arise from both human and technical error during sample collection, preparation, analysis and recording of results [2]. Erroneous values can be recorded even where an organisation has a clearly defined monitoring protocol. If invalid values are subsequently combined with valid data, the integrity of the latter is also impaired [1]. Incorporating erroneous values into a management tool like a WQI or model, could result in wrong conclusions that might be costly to the environment or humans.

Data validation is a rigorous process of reviewing the quality of data. It assists in determining errors and anomalies that might need attention during analysis. Validation is crucial especially where a study depends on secondary data as it increases confidence in the integrity of the obtained data. Without such confidence, further data manipulation is fruitless [3]. Though data validation is usually performed by a quality control personnel in most organisations, it is important for any water resource practitioner to understand the common characteristics that may affect in-depth analysis of a water quality data-sets.

3. Visual scan

Among the common methods of assessing the integrity of a data-set is visual scan. This approach assists to identify values that are distinct and, which might require attention during statistical analysis and model building. The ability to visually assess the integrity of data depends on both the monitoring objectives and experience [4]. Transcription errors, erroneous values (e.g. a pH value of greater than 14, or a negative reading) and inaccurate sample information (e.g. units of mg/L for specific conductivity data) are common errors that can be easily noted by a visual scan. A major source of transcription errors is during data entry or when converting data from one format to another [5, 6]. This is common when data is transferred from a manually recorded spreadsheet to a computer oriented format. The incorrect positioning of a decimal point during data entry is also a common transcription error [7, 8].

A report by [7] suggested that transcription errors can be reduced by minimising the number of times that data is copied before a final report is compiled. [9] recommended the read-aloud technique as an effective way of reducing transcription errors. Data is printed and read-aloud by one individual, while the second individual simultaneously compares the spoken values with the ones on the original sheet. Even though the double data-entry method has been described as an effective method of reducing transcription errors, its main limitation is of being laborious [9-11]. [12], however, recommended slow and careful entry of results as an effective approach of reducing transcription errors.

While it might be easy to detect some of the erroneous values by a general visual scan, more subtle errors, for example outliers, may only be ascertained by statistical methods [13]. Censored values, missing values, seasonality, serial correlation and outliers are common characteristics in data-sets that need identification and treatment [14]. The following sections review the common characteristics in water quality data namely; outliers, missing values and censored values. Methods of their identification and treatment are discussed.

3.1. Outliers (extreme values)

The presence of values that are far smaller or larger than the usual results is a common feature of water quality data. An outlier is defined as a value that has a low probability of originating from the same statistical distribution as the rest of observations in a data-set [15]. Outlying values should be examined to ascertain if they are possibly erroneous. If erroneous, the value can be discarded or corrected, where possible. Extreme values may arise from an imprecise measurement tool, sample contamination, incorrect laboratory analysis technique, mistakes made during data transfer, incorrect statistical distribution assumption or a novel phenomenon, [15, 16]. Since many ecological phenomena (e.g., floods, storms) are known to produce extreme values, their removal assumes that the phenomenon did not occur when actually it did. A decision must thus be made as to whether an outlying datum is an occasional value and an appropriate member of the data-set or whether it should be amended, or excluded from subsequent statistical analyses as it might introduce bias [1].

An outlying value should only be objectively rejected as erroneous after a statistical test indicates that it is not real or when it is desired to make the statistical testing more sensitive [17]. In figure 1, for example, simple inspection might mean that the two spikes are erroneous, but in-depth analysis might correlate the spikes to very poor water quality for those two days, which would make the two observations valid. The model, however, does not pick the extreme values, which negatively affects the R^2 value, and ultimately the accuracy and usefulness of the model in predicting polymer dosage.

Both observational (graphical) and statistical techniques have been applied to identify outliers. Among the common observational methods are the box-plots, time series, histogram, ranked data plots and normal probability plots [18, 19]. These methods basically detect an outlier value by quantifying how far it lies from the other values. This could be the difference between the outlier and the mean of all points, between the outlier and the next closest value or between the outlier and the mean of the remaining values [20].

3.2. Box-plot

The box-plot, a graphical representation of data dispersion, is considered to be a simple observation method for screening outliers. It has been recommended as a primary exploratory tool of identifying outlying values in large data-sets (15). Since the technique basically uses the

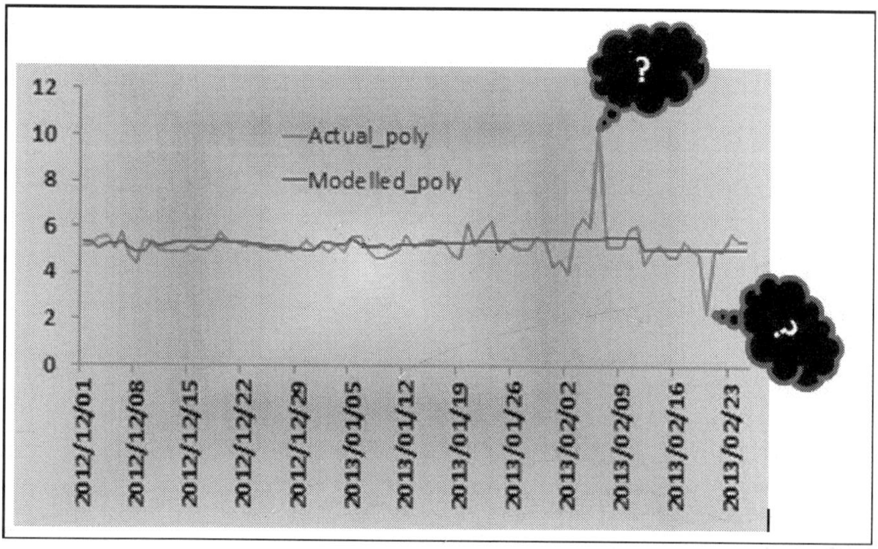

Figure 1. Data inspection during validation and treatment

median value and not the mean, it poses a greater advantage by allowing data analysis disregarding its distribution. [21] and [22] categorised potential outliers using the box-plot as:

- data points between 1.5 and 3 times the Inter Quantile Range (IQR) above the 75th percentile or between 1.5 and 3 times the IQR below the 25th percentile, and
- data points that exceed 3 times the IQR above the 75th percentile or exceed 3 times the IQR below the 25th percentile.

The limitation of a box plot is that it is basically a descriptive method that does not allow for hypothesis testing, and thus cannot determine the significance of a potential outlier [15].

3.3. Normal probability plot

The probability plot method identifies outliers as values that do not closely fit a normal distribution curve. The points located along the probability plot line represent 'normal',observation, while those at the upper or lower extreme of the line, indicates the suspected outliers as depicted in Figure 2.

The approach assumes that if an extreme value is removed, the resulting population becomes normally distributed [21]. If, however, the data still does not appear normally distributed after the removal of outlying values, a researcher might have to consider normalising it by transformation techniques, such as using logarithms [21, 23]. However, it should be highlighted that data transformation tends to shrink large values (see the two extreme values in Figure 1, before transformation), thus suppressing their effect which might be of interest for further

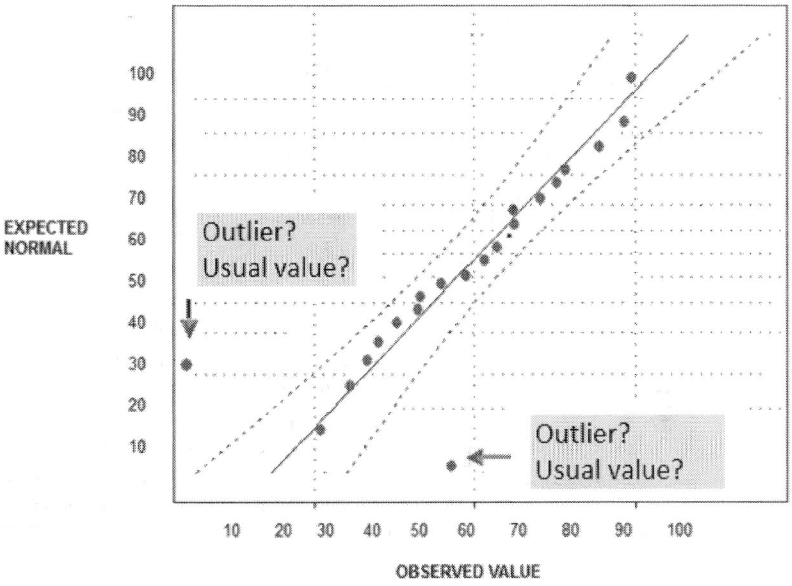

Figure 2. Normal probability plot showing outliers

analysis [23, 24]. Data should thus not be simply transformed for the sole purpose of eliminating or reducing the impact of outliers. Furthermore, since some data transformation techniques require non-negative values only (e.g. square root function) and a value greater than zero (e.g. logarithm function), transformation should not be considered as an automatic way of reducing the effect of outliers [23].

Since observational methods might fail to identify some of the subtle outliers, statistical tests may be performed to identify a data point as an outlier. However a decision still has to be made on whether to exclude or retain an outlying data point. The section below describes the common statistical test for identifying outliers.

3.4. Grubbs test

The Grubb's test, also known as the Studentised Deviate test, compares outlying data points with the average and standard deviation of a data-set [25-27]. Before applying the Grubbs test, one should firstly verify that the data can be reasonably approximated by a normal distribution. The test detects and removes one outlier at a time until all are removed. The test is two sided as shown in the two equations below.

1. To test whether the maximum value is an outlier, the test:

$$G_{max} = \frac{X_n - X_{mean}}{S}$$

2. To test whether the minimum value is an outlier, the test is:

$$G_{min} = \frac{X_{mean} - X_i}{S}$$

Where X_1 or X_n=the suspected single outlier (max or min)

s=standard deviation of the whole data set

X=mean

The main limitation of Grubbs test is of being invalid when data assumes non-normal distribution [28]. Multiple iterations of data also tends to change the probabilities of detection. Grubbs test is only recommended for sample sizes of not more than six, since it frequently tags most of the points as outliers. It suffers from masking, which is failure to identify more than one outlier in a data-set [28, 29]. For instance, for a data-set consisting of the following points; 3, 5, 7, 13, 15, 150, 153, the identification of 153 (maximum value) as an outlier might fail because it is not extreme with respect to the next highest value (150). However, it is clear that both values (150 and 153) are much higher than the rest of the data-set and could jointly be considered as outliers.

3.5. Dixon test

Dixon's test is considered an effective technique of identifying an outlier in a data-set containing not more than 25 values [21, 30]. It is based on the ratio of the ranges of a potential outlier to the range of the whole data set as shown in equation 1 [31]. The observations are arranged in ascending order and if the distance between the potential outlier to its nearest value (Q_{gap}) is large enough, relative to the range of all values (Q_{range}), the value is considered an outlier.

$$Q_{exp} = \frac{Q_{gap}}{Q_{range}} \quad (1)$$

The calculated Q_{exp} value is then compared to a critical Q-value (Qcrit) found in tables. If Q_{exp} is greater than the suspect value, the suspected value can be characterised as an outlier. Since the Dixon test is based on ordered statistics, it tends to counter-act the normality assumption [15]. The test assumes that if the suspected outlier is removed, the data becomes normally distributed. However, Dixon's test also suffers the masking effect when the population contains more than one outlier.

[32] recommended the use of multivariate techniques like Jackknife distance and Mahalanobis distance [33, 34]. The strength of multivariate methods is on their ability to incorporation of the correlation or covariance between variables thus making them more correct as compared to univariate methods. [34] introduced the chi-square plot, which draws the empirical distribution function of the robust Mahalanobis distances against the chi-square distribution. A value that is out of distribution tail indicates that it is an outlier [33].

For an on-going study, an outlier can be ascertained by re-analysis of the sample, if still available and valid. [28] and [2] advised the practise of triplicate sampling as an effective method of verifying the unexpected results. When conducting a long-term study, researchers might consider re-sampling when almost similar conditions prevail again. Nevertheless, this option might not be feasible when carrying out a retrospective study since it generally depend on secondary data from past events.

For data intended for trend analysis, studies have recommended the application of nonparametric techniques such as the Seasonal Kendal test where transformation techniques do not yield symmetric data [19]. Should a parametric test be preferred on a data-set that includes outliers, practitioners may evaluate the influence of outliers by performing the test twice, once using the full data-set (including the outliers) and again on the reduced data-set (excluding the outliers). If the conclusions are essentially the same, then the suspect datum may be retained, failing which a nonparametric test is recommended.

4. Missing values

While most statistical methods presumes a complete data-set for analysis, missing values are frequently encountered problems in water quality studies [35, 36]. Handling missing values can be a challenge as it requires a careful examination of the data to identify the type and pattern of missingness, and also have a clear understanding of the most appropriate imputation method. Gaps in water quality data-sets may arise due to several reasons, among which are imperfect data entry, equipment error, loss of sample before analysis and incorrect measurements [37]. Missing values complicate data analysis, cause loss of statistical efficiency and reduces statistical estimation power [37-39]. For data intended for time-series analysis and model building, gaps become a significant obstacle since both generally require continuous data [40, 41]. Any estimation of missing values should be done in a manner that minimise the introduction of more bias in order to preserve the structure of original data-set [41, 42].

The best way to estimate missing values is to repeat the experiment and produce a complete data-set. This option is however, not feasible when conducting a retrospective study since it depend on historical data. Where it is not possible to re-sample, a model or non-model techniques may be applied to estimate missing values [43].

If the proportion of missing values is relatively small, listwise deletion has been recommended. This approach, which is considered the easiest and simplest, discards the entire case where any of the variables are missing. Its major advantage is that it produce a complete data-set, which in turn allows for the use of standard analysis techniques [44]. The method also does not require special computational techniques. However, as the proportion of missing data increases, deletion tends to introduce biasness and inaccuracies in subsequent analyses. This tends to reduce the power of significance test and is more pronounced particularly if the pattern of missing data is not completely random. Furthermore, listwise deletion also decreases the sample size which tends to reduce the ability to detect a true association. For example, suppose a data-set with 1,000 samples and 20 variables and each of the variables has missing data for 5% of the cases, then, one could expect to have complete data for only about 360 individuals, thus discarding the other 640.

On the other hand, pairwise deletion removes incomplete cases on an analysis-by-analysis basis, such that any given case may contribute to some analyses but not to others [44]. This approach is considered an improvement over listwise deletion because it minimises the number of cases discarded in any given analysis. However, it also tend to produce bias if the data is not completely random.

Several studies have applied imputation techniques to estimate missing values. A common assumption with these methods is that data should be missing randomly [45]. The most common and easiest imputation technique is replacing the missing values with an arithmetic mean for the rest of the data [35, 41]. This is recommended where the frequency distribution of a variable is reasonably symmetric, or has been made so by data transformation methods. The advantage of arithmetic mean imputation is generation of unbiased estimates if the data is completely random because the mean lands on the regression line. Even though the insertion of mean value does not add information, it tends to improve subsequent analysis. However, while simple to execute, this method does not take into consideration the subjects patterns of scores across all the other variables. It changes the distribution of the original data by narrowing the variance [46]. If the data assumes an asymmetric distribution, the median has been recommended as a more representative estimate of the central tendency and should be used instead of the mean.

[47], recommended model-based substitution techniques as more flexible and less ad hoc approach of estimating missing values as compared to non-model methods. A simple modelling technique is to regress the previous observations into an equation which estimates missing values [35, 48]. The time-series auto-regressive model has been described as an improvement and more accurate method of estimating missing values [25]. Unlike the arithmetic mean and median replacement methods, regression imputation techniques estimates missing values of a given variable using data of other parameters. This tends to reduce the variance problem, which is common with the arithmetic mean imputation and median replacement methods [41, 49].

On the other hand, the maximum likelihood technique uses all the available complete and incomplete data to identify the parameter values that have the highest probability of producing the sample data [44]. It runs a series of data iterations by replacing different values for the unknown parameters and converges to a single set of parameters with the highest probability of matching the observed data [41]. The method has been recommended as it tends to give efficient estimates with correct standard errors. However, just like other imputation methods, the maximum likelihood estimates can be heavily biased if the sample size is small. In addition, the technique requires a specialised software which may be expensive, challenging to use and time consuming.

Some studies have considered the relationship between parameters as an effective approach of estimating missing values [50]. For instance, missing conductivity values can be calculated from the total dissolved solids value (TDS) by a simple linear regression where p-value and r-value are known to exist and the missing value lies between the two variables. Equation 2, where ***a*** is in the range 1.2-1.8, has been described as an equally important estimator of missing conductivity values [1, 51].

$$Conductivity \approx TDS \ x \ a \tag{2}$$

The constant, *a*, is high in water of high chloride and low sulphate [51]. [52] estimated missing potassium values by using a linear relationship between potassium and sodium. The relationship gave a high correlation coefficient of 0.904 ($p<0.001$).

As of late, research has explored the application of artificial intelligence (AI) techniques to handle missing values in the water quality sector. Among the major AI techniques that have been applied is the Artificial Neural Networks (ANN) and Hybrid Evolutionary Algorithms (HEA) (48, 53, 54). Nevertheless, it should also be highlighted that all techniques for estimating missing values invariably affect the results. This is more pronounced when missing values characterise a significant proportion of the data being analysed. A research should thus consider the sample size when choosing the most appropriate imputation method.

5. Scientific facts

The integrity of water quality data can also be assessed by checking whether the results are inline with known scientific facts. To ascertain that, a researcher must have some scientific knowledge regarding the characteristics of water quality variables. Below are some scientific facts that can be used to assess data integrity [1].

1. Presence of nitrate in the absence of dissolved oxygen may indicate an error since nitrate is rapidly reduced in the absence of oxygen. The dissolved oxygen meter might have malfunctioned or oxygen might have escaped from the sample before analysis.

2. Component parts of a water-quality variable must not be greater than the total variable. For example:

3. Total phosphorus ≥Total dissolved phosphorus>Ortho-phosphate.

4. Total Kjeldahl nitrogen ≥Total dissolved Kjeldahl nitrogen>ammonia.

5. Total organic carbon ≥Dissolved organic carbon.

6. Species in a water body should be described correctly with regards to original pH of the water sample. For example, carbonate species will normally exist as HCO_3^- while CO_3^{2-} cannot co-exist with H_2CO_3.

6. Censored data

A common problem faced by researchers analysing environmental data is the presence of observations reported to have non-detectable levels of a contaminant. Data which are either less than the lower detection limit, or greater than the upper detection limit of the analytical method applied are normally artificially curtailed at the end of a distribution, and are termed "censored values" [14]. Multiple censored results may be recorded when the laboratory has changed levels of detection, possibly as a result of an instrument having gained more accuracy, or the laboratory protocol having established new limits. If the values are below the detection limit, they are abbreviated as BDL, and when above the limit, as ADL [55, 56].

Various methods of treating censored values have been developed to reduce the complication generally brought about by censored values [57]. The application of an incorrect method may introduce bias especially when estimating the mean and variance of data distribution [58]. This may consequently distort the regression coefficients and their standard errors, and further reduce the hypothesis testing power. A researcher must thus decide on the most appropriate method to analyse censored values. One might reason that since these values are extra ordinarily small, they are not important and discard them while some might be tempted to remove them inorder to ease statistical analysis. Deletion has however been described as the worst practise as it tends to introduce a strong upward bias of the central tendency which lead to inaccurate interpretation of data [19, 59-62].

The relatively easiest and most common method of handling censored values is to replace them with a real number value so that they conform to the rest of data. The United State Environmental Agency suggested substitution if censored data is less than 15% of the total data-set (63, 64). [8], BDL, for example $x < 1.1$, were multiplied by the factor 0.75 to give 0.825. ADL values, for example $500 < x$, were recorded as one magnitude higher than the limit values to give 501. [65] recommended substituting with $\frac{1}{2}$ DL or $\frac{1}{\sqrt{2}}$ DL if the sample size is less than 20 and contains less than 45% of its data as censored values. [66] suggested substitution by $\frac{1}{\sqrt{2}}$DLif the data are not highly skewed and substitution by $\frac{1}{2}$ DL otherwise. [67], however, criticised the substitution approach and illustrated how the practice could produces poor estimates of correlation coefficients and regression slopes. [68] further explained that substitution is not suitable if the data has multiple detection limits [68, 69].

A second approach for handling censored values is the maximum likelihood estimation (MLE). It is recommended for a large data-set which assumes normality and contains censored results [38, 65, 70, 71]. This approach basically uses the statistical properties of non-censored portion of the data-set, and an iterative process to determine the means and variance. The MLE technique generates an equation that calculates mean and standard deviation from values assumed to represent both the detects and non-detect results [69]. The equation can be used to estimate values that can replace censored values. However, the technique is reportedly ineffective for a small data-set that has fewer than 50 BDLs [69].

When data assumes an independent distribution and contain censored values, non-parametric methods like the Kaplan-Meir method, can be considered for analysis [59]. The Kaplan-Meir method creates an estimate of the population mean and standard deviation, which is adjusted for data censoring, based on the fitted distribution model. Just like any non-parametric techniques for analysing censored data, the Kaplan-Meier is only applicable to right-censored results (i.e. greater than) [72]. To use Kaplan-Meier on left-censored values, the censored values must be converted to right-censored by flipping them over to the largest observed value [65, 71, 72]. To ease the process, [73] have developed a computer program that does the conversion. [71], however, found the Kaplan-Meier method to be effective when summarising a data-set containing up to 70% of censored results.

In between the parametric and non-parametric methods is a robust technique called Regression on Order Statistics (ROS) [38]. It treats BDLs based on the probability plot of detects. The technique is applicable where the response variable (concentration) is a linear function of the explanatory variable (the normal quartiles) and if the error variance of the model is constant. It also assumes that all censoring thresholds are left-censored and is effective for a data-set which contains up to 80% censored values [59]. The ROS technique uses data plots on a modelling distribution to predict censored values. [59] and [68] evaluated ROS as a reliable method for summarising multiply censored data. Helsel and Cohn (38) also described ROS as a better estimator of the mean and standard deviation as compared to MLE, when the sample size is less than 50 and contains censored values.

7. Statistical methods

The success of an analysis of water quality data primarily depends on the selection of the right statistical method which considers common data characteristics such as normality, seasonality, outliers, missing values, censoring, etc., [74]. If the data assumes an understandable and describable distribution, parametric methods can be used [14]. However, non-parametric techniques are slowly replacing parametric techniques mainly because the latter are sensitive to common water characteristics like outliers, missing values and censored value [75].

7.1. Computer application in data treatment

The increase in various computer programs has made it easy to detect and treat erroneous data. Computers now provide flexibility and speedy methods of data analysis, tabulation, graph preparation or running models, among others. Various software such as Microsoft Excel, Minitab, Stata and MATLAB have become indispensable tools for analysing environmental data. These software perform various computations associated with checking assumptions about statistical distributions, error detection and their treatment. However, the major problem encountered by researchers, is lack of guidance regarding selection of the most appropriate software. Computer-aided statistical analysis should be undertaken with some understanding of the techniques being used. For example, some statistical software packages might replace missing values with the means of the variable, or prompt the user for case-wise deletion of analytical data, both of which might be considered undesirable [52].

Lately, machine learning algorithms like the artificial neural networks (ANNs) [67, 76-78], and genetic algorithms (GA) [76, 79] have gained momentum in water quality monitoring studies. [41] pointed out that these technique generally yields the best parameter estimates in the data set with the least amount of missing data. Nevertheless, as the percentage of missing data increases, the performance of ANN which is generally measured by the errors in the parameter estimates, decreases and may reach performance levels similar to those obtained by the general substitution methods. However, in all cases the effectiveness of these methods lies on the user's ability to manipulate and display data correctly.

8. Conclusion

This chapter discussed the common data characteristics which tend to affect statistical analysis. It is recommended that practitioners should explore for outliers, missing values and censored values in a data-set before undertaking in-depth analysis. Although an analyst might not be able to establish the causal of such characteristic, eliminate or overcome some of the errors, having knowledge of their existence assists in establishing some level of confidence in drawing meaningful conclusions. It is recommended that water quality monitoring programs should strive to collect data of high quality. Common methods of ascertaining data quality are practising duplicate samples, using blanks or reference samples, and running performance audits. If a researcher is not sure of how to treat a characteristic of interest, a non-parametric method like Seasonal Kendal test could provide a better alternative since it is insensitive to common water quality data characteristics like outliers.

Author details

Innocent Rangeti[1], Bloodless Dzwairo[2], Graham J. Barratt[1] and Fredrick A.O. Otieno[3]

*Address all correspondence to: innoranger@gmail.com

1 Department of Environmental Health, Durban University of Technology, Durban, South Africa

2 Institute for Water and Wastewater Technology, Durban University of Technology, Durban, South Africa

3 DVC: Technology, Innovation and Partnerships, Durban University of Technology, Durban, South Africa

References

[1] Steel A, Clarke M, Whitfield P. Use and Reporting of Monitoring Data. In: Bartram J, Ballance R, editors. Water Quality Monitoring-A Practical Guide to the Design and Implementation of Freshwater Quality Studies and Monitoring Programmes: United Nations Environment Programme and the World Health Organization; 1996.

[2] Mitchell P. Guidelines for Quality Assurance and Quality Control in Surface Water Quality Programs in Alberta. Alberta Environment, 2006.

[3] Tasić S, Feruh MB. Errors and Issues in Secondary Data used in marketing research. The Scientific Journal for Theory and Practice of Socioeconomic Development. 2012;1(2).

[4] Doong DJ, Chen SH, Kao CC, Lee BC, Yeh SP. Data quality check procedures of an operational coastal ocean monitoring network. Ocean Engineering. 2007;34(2):234-46.

[5] Taylor S, Bogdan R. Introduction to research methods: New York: Wiley; 1984.

[6] Wahi MM, Parks DV, Skeate RC, Goldin SB. Reducing errors from the electronic transcription of data collected on paper forms: a research data case study. Journal of the American Medical Informatics Association. 2008;15(3):386-9.

[7] UNEP-WHO. Use and Reporting of monitoring data. In: Bartram J, Ballace R, editors. Water Quality Monitoring A practical Guide to the Design and Implementation of Freshwater Quality Studies and Monitoring Programmes. London: United Nations Environmental program, World Health Organisation; 1996.

[8] Dzwairo B. Modelling raw water quality variability in order to predict cost of water treatment. Pretoria: Tshwane University of Technology; 2011.

[9] Kawado M, Hinotsu S, Matsuyama Y, Yamaguchi T, Hashimoto S, Ohashi Y. A comparison of error detection rates between the reading aloud method and the double data entry method. Controlled Clinical Trials. 24. 2003:560–9.

[10] Cummings J, Masten J. Customized dual data entry for computerized analysis. Quality Assurance: Good Practice, Regulation, and Law. 1994;3:300-3.

[11] Brown ML, Austen DJ. Data management and statistical techniques. Murphy BR, Willis DW, editors. Bethesda, Maryland: America Fisheries Society; 1996.

[12] Rajaraman V. Self study guide to Analysis and Design of Information Systems. New Delhi: Asoke K. Ghosh; 2006.

[13] Data Analysis and Interpretation. Rhe Monitoring Guideline. Austria2000.

[14] Helsel DR, Hirsch RM. Statistical Methods in Water Resources. Amsterdam, Netherlands: Elsevier Science Publisher B.V; 1992.

[15] Köster D, Hutchinson N. Review of Long-Term Water Quality Data for the Lake System Health Program. Ontario: 2008 Contract No.: GLL 80398.

[16] Iglewicz B, Hoaglin DC. How to Detect and Handle Outliers1993.

[17] Zar JH. Biostatistical Analysis. Upper Saddle River, NJ.: Prentice-Hall Inc; 1996.

[18] Silva-Ramírez E-L, Pino-Mejías R, López-Coello M, Cubiles-de-la-Vega M-D. Missing value imputation on missing completely at random data using multilayer perceptrons. Neural Networks. 2011;24(1):121-9.

[19] USEPA., Ecology. Do. Technical Guidance for Exploring TMDL Effectiveness Monitoring Data. 2011.

[20] Stoimenova E, Mateev P, Dobreva M. Outlier detection as a method for knowledge extraction from digital resources. Review of the National Center for Digitization. 2006;9:1-11.

[21] US-EPA. Statistical Analysis of Groundwater Monitoring Data at RCRA Facilities, Unified Guidance (Unified Guidance). US-Environmetal Protection Agency, 2009.

[22] US-EPA. Data Quality Assessment: Statistical Methods for Practitioners. In: Agency USEP, editor: United States Environmental Protection Agency; 2006.

[23] Osoborne JW. Improving your data transformations: Applying the Box-Cox transformation. Practical Assessment, Research and Evaluation. 2010;15(12).

[24] High R. Dealing with 'Outlier': How to Maintain your data's integrity.

[25] Chi Fung DS. Methods for the Estimation of Missing Values in Time Series. Western Australia: Edith Cowan University; 2006.

[26] US-EPA. Statistical Training Course for Ground-Water Monitoring Data Analysis.. Washington D.C.: Environmental Protection Agency, 1992.

[27] Grubbs FE, Beck G. Extension of sample sizes and percentage points for significance tests of outlying observations. Technometric. 1972;14:847-54.

[28] Tiwari RC, Dienes TP. The Kalman filter model and Bayesian outlier detection for time series analysis of BOD data. Ecological Modelling 1994;73:159-65.

[29] De Muth JE. Basic statistics and pharmaceutical statistical applications: CRC Press; 2014.

[30] Gibbons RD. Statistical Methods for Groundwater Monitoring. New York: John Wiley & Sons; 1994.

[31] Walfish S. A Review of Statistical Outlier Methods. Pharmaceutical Technology. 2006.

[32] Robinson RB, Cox CD, Odom K. Identifying Outliers in Correlated Water Quality Data. Journal of Environmental Engineering. 2005;131(4):651-7.

[33] Filzmoser P, editor A multivariate outlier detection method. Proceedings of the seventh international conference on computer data analysis and modeling; 2004: Minsk: Belarusian State University.

[34] Garret D. The Chi-square plot: A tool for multivariate recognition. Journal of Geochemical Exploration. 1989;32:319-41.

[35] Ssali G, Marwala T. Estimation of Missing Data Using Computational Intelligence and Decision Trees. n.d.

[36] Noor NM, Zainudin ML. A Review: Missing Values in Environmental Data Sets. International Conference on Environment 2008 (ICENV 2008); Pulau Pinang2008.

[37] Calcagno G, Staiano A, Fortunato G, Brescia-Morra V, Salvatore E, Liguori R, et al. A multilayer perceptron neural network-based approach for the identification of responsiveness to interferon therapy in multiple sclerosis patients. Information Sciences. 2010;180(21):4153-63.

[38] Helsel DR, Cohn T. Estimation of description statistics for multiply censored water quality data. Water Resource Research. 1988; 24:1997-2004.

[39] Little RJ, Rubin DB. Statistical analysis with missing data. New York: John Wiley and Sons; 1987.

[40] Junninen H, Niska H, Tuppurainen K, Ruuskanen J, Kolehmainen M. Methods for imputation of missing values in air quality data sets. Atmospheric Environment. 2004; 38:2895–907.

[41] Nieh C. Using Mass Balance, Factor Analysis, and Multiple Imputation to Assess Health Effects of Water Quality. Chicago, Illinois: University of Illinois 2011.

[42] Luengo J, Garcia S, Herrera F. A study on the use of imputation methods for experimentation with Radial Basis Function Network classifiers handling missing attribute values: The good synergy between RBFNs and EventCovering methodI. Neural Networks. 2010; 23:406-18.

[43] Lakshminarayan K, Harp S, Samad T. Imputation of missing data in industrial databases. Applied Intelligence. 1999;11(3):259-75.

[44] Baraldi AN, Enders CK. An introduction to modern missing data analyses. Journal of School Psychology. 2010;48(1):5-37.

[45] Rubin DB. Inference and missing data. Biometrika. 1976;63(3):581-92.

[46] Enders CK. A primer on the use of modern missing-data methods in psychosomatic medicine research. PsychosomaticMedicine. 2006;68:427–36.

[47] Fogarty DJ. Multiple imputation as a missing data approach to reject inference on consumer credit scoring. Intersat. 2006.

[48] Smits A, Baggelaar PK. Estimating missing values in time series. Netherlands: Association of River Waterworks – RIWA; 2010.

[49] Sartori N, Salvan A, Thomaseth K. Multiple imputation of missing values in a cancer mortality analysis with estimated exposure dose. ComputationalStatistics and DataAnalysis. 2005;49(3):937–53.

[50] Güler C, Thyne GD, McCray JE, Turner AK. Evaluation of graphical and multivariate statistical methods for classification of water chemistry data. Hydrogeology 2002;10:455–74.

[51] Dzwairo B, Otieno FAO, Ochieng' GM. Incorporating surface raw water quality into the cost chain for water services: Vaal catchment, South Africa. Research Journal of Chemistry and Environment. 2010;14(1):29-35.

[52] Guler C, Thyne GD, McCray JE, Turner AK. Evaluation of graphical and multivariate statistical methods for classification of water chemistry data. Hydrogeology. 2002;10:455-74.

[53] Starret KS, Starret KS, Heier T, Su Y, Tuan D, Bandurraga M. Filling in Missing Peakflow data using Artificial Neural Networks. Journal of Engineering and Applied Science. 2010;5(1).

[54] Aitkenhead MJ, Coull MC. An application based on neural networks for replacing missing data in large datasets n.d.

[55] Lin P-E, Niu X-F. Comparison of Statistical Methods In Handling Minimum Detection Limits. Department of Statistics,, Florida State University, 1998.

[56] Darken PF. Testing for changes in trend in water quality data. Blacksburg, Virginia: Virginia Polytechnic Institute and State University; 1999.

[57] Farnham IM, Stetzenbach KJ, Singh AS, Johannesson KH. Treatment of nondetects in multivariate analysis of groundwatervgeochemistry data. Chemometrics Intelligent Lab Sys. 2002;60(265-281).

[58] Lyles RH, Fan D, Chuachoowong R. Correlation coefficient estimation involving a left censored laboratory assay variable. Statist Med. 2001;20:2921-33.

[59] Lopaka L, Helsel D. Statistical analysis of water-quality data containing multiple detection limits: S-language software for regression on order statistics. Computers & Geosciences. 2005;31:1241-8.

[60] Gheyas IA, Smith LS. A neural network-based framework for the reconstruction of incomplete data sets. Neurocomputing. 2010;73(16–18):3039-65.

[61] Nishanth KJ, Ravi V, Ankaiah N, Bose I. Soft computing based imputation and hybrid data and text mining: The case of predicting the severity of phishing alerts. Expert Systems with Applications. 2012;39(12):10583-9.

[62] Helsel D. Fabricating data: How substituting values for nondetects can ruin results, and what can be done about it. Chemosphere. 2006;65:2434-9.

[63] Kalderstam J, Edén P, Bendahl P-O, Strand C, Fernö M, Ohlsson M. Training artificial neural networks directly on the concordance index for censored data using genetic algorithms. Artificial Intelligence in Medicine. 2013;58(2):125-32.

[64] Environmental Protection Agency. Data Quality Assessment: Statistical Methods for Practitioners. Agency USEP, editor. Washington: United States Environmental Protection Agency, Office of Environmental Information, 2006; 2006 14 June 2013. 198 p.

[65] Hewett P, Ganser GH. A Comparison of Several Methods for Analyzing Censored Data. British Occupational Hygiene. 2007;57(7):611-32.

[66] Hornung RW, Reed LD. Estimation of Average Concentration in the Presence of Nondetectable Values. Applied Occupational and Environmental Hygiene. 1990;5(1): 46-51.

[67] Kang P. Locally linear reconstruction based missing value imputation for supervised learning. Neurocomputing. 2013(0).

[68] Shumway RH, Azari RS, Kayhanian M. Statistical approaches to estimating mean water quality concentrations with detection limits. Environmental Science and Technology. 2002;36:3345-53.

[69] Helsel D. Less than obvious-statistical treatment data below the detection limit. Environmental Science and Technology. 1990;24(12):1766-44.

[70] Sanford RF, Pierson CT, Crovelli RA. An objective replacement method for censored geochemical data. Math Geol. 1993;25:59-80.

[71] Antweiler RC, Taylor HE. Evaluation of Statistical Treatments of Left-Censored Environmental Data using Coincident Uncensored Data Sets. Environmental Science and Technology. 2008;42(10):3732-8.

[72] Fu L, Wang Y-G. Statistical tools for analysing water quality data2012.

[73] Silva JdA, Hruschka ER. An experimental study on the use of nearest neighbor-based imputation algorithms for classification tasks. Data & Knowledge Engineering. 2013;84(0):47-58.

[74] Visser A, Dubus I, Broers HP, Brouyere S, Korcz M, Orban P, et al. Comparison of methods for the detection and extrapolation of trends in groundwater quality. Journal of Environmental Monitoring. 2009;11(11):2030-43.

[75] Schertzer TL, Alexander RB, Ohe DJ. The Computer Program Estimate Trend (ESTREND), a system for the detection of trends in water-quality data Water Resources investigation report. 1991;91-4040:56-7.

[76] Recknagel F, Bobbin J, Whigham P, Wilson H. Comparative application of artificial neural networks and genetic algorithms for multivariate time-series modelling of algal blooms in freshwater lakes. Journal of Hydroinformatics. 2002;4(2):125-34.

[77] Lee JHW, Huang Y, Dickman M, Jayawardena AW. Neural network modelling of coastal algal blooms.Ecological Modelling. 2003;159(179-201).

[78] Singh KP, Basant A, Malik A, Jain G. Artificial neural network modeling of the river water quality—a case study. Ecological Modelling. 2009;220(6):888-95.

[79] Muttil N, Lee JHW. Genetic programming for analysis and real-time prediction of coastal algal blooms. Ecological Modelling. 2005;189:363-76.

Chapter 5

Quality of Water Quality Data — Consistency of the Results of Chemical Analyses and Sources of Uncertainty

Nataša Gros

Additional information is available at the end of the chapter

http://dx.doi.org/10.5772/59012

1. Introduction

Water quality has several aspects and brings together professionals with different backgrounds, which have diverse roles, participate and contribute in their own way and look at water quality issues from different perspectives. One of possible distinctions of roles is that on one side we have the providers of results of chemical analyses and on the other the users of these results. The providers, analytical chemists, have a good understanding of an analytical process, knowledge on the abilities and limitations of analytical methods and expertise in using them. The users of results, water quality experts and hydro-geologists have proficiency in analysing and explaining results within a wider context, drawing conclusions on the water quality, recognising trends and processes within a water body.

Chemical analyses can have different scopes and consequently different levels of complexity [1]. They can be either extended, tending to be as complete as possible, or partial, comprising only a selected number of chemical parameters, and scope or use orientated. Examples of the later are analyses intended for considerations of eutrophication processes in a water body, contamination of water with pesticides, or water suitability for irrigation or recreation.

The consistency and reliability of the results of chemical analyses are as essential for analytical chemists as they are for water experts. Analytical chemists on one side need to keep their professional reputation. Water experts on the other have to draw relevant conclusions from the provided results. Each side has different level of insight in to quality of the provided data, which they understand and treat differently. Water experts have rather limited options if an analysis is only partial and narrowly scoped. Their judgement is more or less limited to the consideration how likely the results seem to be within an established context, taking into account hypotheses, expectations and previous experiences. With an extended water analysis the situation is very different and water experts have at their disposal several tests and miscellaneous checks on the consistency and coherency of analytical results [2].

In an extended water analysis a total ionic composition of a water sample comprising anions and cations is known. A consideration of a cation-anion balance is consequently the most basic test which can be applied. If an imbalance does not exceed 5 % it can be expected that an analysis as a whole comprises all main ions and contains no grave errors. Such consideration is in a way some sort of a top down approach telling nothing on the uncertainties of individual results nor taking them into account. The acceptable imbalance limit of 5 % is an agreed arbitrary value derived from experiences. Good agreement between anionic and cationic composition in terms of their balance may not be enough proof of the completeness of analysis and the calculated parameter (salinity×1000)/TDS seems to be a useful aid for reliable judgement for seawater and derived saline solutions as we demonstrated [3]. TDS stands for total dissolved solids usually expressed in mg/L. The judgement on the coherence of analytical results can be further supported by miscellaneous checks, considering additional parameters such as electric conductivity, pH, total dissolved solids, a dry residual and water density [2]. Evaluation of the proportions between concentrations of ions also applies. The judgement is based on a consideration if the established proportions are naturally possible and in a good agreement with what is known on the origin of water or its genesis.

An approach used by analytical chemists in evaluating the quality of analytical results is different. Analytical chemists limit errors in analytical results by gaining control over an analytical process. They can achieve this by having a system of quality assurance (QA) and quality control (QC) implemented into laboratory practice [4]. For an analytical chemist to choose an analytical method that is fit for purpose it is essential to understand the needs of a user of analytical results well. The method should be validated and under an internal and preferably also external quality control to produce reliable results. By considering critically all steps of an analytical process analytical chemists recognise sources of uncertainty and estimate a combined standard uncertainty of an analytical result by following the theory on the propagation of errors. By stating uncertainty they give to the user some additional insight into reliability of an analytical result. The approach described here is called a bottom-up approach in evaluating uncertainty. There are several papers and recommendations of official bodies which give guidance in evaluation of uncertainty of analytical results [5-8]. For a reader not specialised in analytical chemistry we give in next few paragraphs a brief account of the methodology used in quantifying uncertainty in analytical chemistry.

An uncertainty evaluation with a bottom-up approach is based on a detailed and comprehensive evaluation of a total analytical process. When considering the process a visualisation of its major steps in a form of a flow chart is usually helpful in recognising the sources of uncertainty. The next step is to write a mathematical equation, a model equation which defines the final result of an analysis y and relates this result to measured quantities x_i. By doing this we bring to light the uncertainty contributions, which we can further analyse in depth by drawing a cause and effect diagram, called also a fish bone diagram. The importance of this step is that it can prevent unjustifiable double counting of the same contribution. Another advantage is that it can lead to a simplification of a model and recognition that some uncer-

http://dx.doi.org/10.5772/59012

tainty sources in fact do not contribute and can be omitted. The model refined in such a way is a starting point for an estimation of the combined standard uncertainty $u_c(y)$ associated with the final result of an analysis. This estimation requires that all the uncertainty contributions are expressed as standard uncertainties $u(x_i)$ corresponding to standard deviations, and that they are combined by following the theory of propagation of errors.

If the uncertainties are primarily not given in such a way they should be recalculated into $u(x_i)$. Such examples are uncertainties which are expressed as a range ($\pm a$) or given in a form of tolerance limits. A range is converted into $u(x_i)$ by taking into account an appropriate distribution function. The distributions most frequently considered are a rectangular or a triangular distribution to which the equations $u(x_i)=a/\sqrt{3}$ or $u(x_i)=a/\sqrt{6}$ apply.

The theory on propagation of errors suggests how standard uncertainties should be combined considering mathematical operations used in a model equation. For the mathematical operations such as summation, subtraction, division and multiplication that are most frequently used in model equations we summarise the basic rules here. For more advanced cases guidance can be found in literature. In the theory on propagation of errors the mathematical operations of summation and subtraction are treated equally. The same is true for multiplication and division. With a model equation $y=x_1+x_2-x_3$ the equation for calculating the combined standard uncertainty of the result is $u_c(y)=(u(x_1)^2+u(x_2)^2+u(x_3)^2)^{1/2}$, and with a model equation $y=x_1 \times x_2 / x_3$ the equation for calculating the combined standard uncertainty reads $u_c(y)=y \times (u(x_1)^2/x_1^2+u(x_2)^2/x_2^2+u(x_3)^2/x_3^2)^{1/2}$. In different words in the first case the variances or the squares of standard uncertainties are combined; and in the second case the squares of relative standard uncertainties are combined.

It has to be noted that not all uncertainty sources necessarily directly reflect in a model equation. Such an example is inhomogeneity of a sample or uncertainty in concentration of a stock calibration standard solution. Such contributions are taken into account at the end and combined with the standard uncertainty obtained previously by combining the uncertainties related to the model equation. The law on propagation of errors for mathematical operations of summation is applied.

It is common practice that the final result y is stated with the expanded standard uncertainty $U(y)$, namely $y \pm U(y)$. The expanded uncertainty is calculated by the equation $U(y)=k \times u_c(y)$; and k stands for a coverage factor. The value of k should always be stated. It is usually between 2 and 3, but most frequently 2. With the k value 2 and a normal distribution of results it can be expected with approximately 95 % level of confidence that the measurand Y lies within the interval $y \pm U(y)$.

We briefly explained methodology for evaluating uncertainty of analytical results by using algebraic equations. But more in depth uncertainty evaluations tend not only to estimate the expanded uncertainty of the final result but to analyse the uncertainty contributions to the combined standard uncertainty. By doing this analytical chemists gain understanding of the major contributions to the combined standard uncertainty and can make effort in improving analytical process by reducing the major uncertainty sources. The results of such more in depth uncertainty evaluation are usually presented in a form of a bar graph in which the combined standard uncertainty is presented in the comparison with the uncertainty contributions of measured quantities x_i. Such a graph can be obtained by a spreadsheet calculation method suggested by Kragten [7, 9] and easily performed with MS Excel software. Another numerical method which is gaining more and more attention in uncertainty evaluations is Monte Carlo simulation; and Chew demonstrated that a spreadsheet calculation method can also be applied to it [10].

For routine ion chromatographic analyses of drinking water of a low mineralisation a validation process is well established and sources of uncertainty well defined. For more versatile and highly mineralised water samples [11] with a complex matrix a situation is different and recognition and judgement of all uncertainty sources is far away from being trivial, well established and resolved. And this is the area where this chapter tends to make a contribution.

Ion chromatography is the most universal and the farthest reaching single method for determining the total ionic composition of natural waters [12-14]. Total ionic composition of a sample can be determined with ion chromatography at least predominantly if not completely, and this can usually be achieved within two separate runs, one for cations and another for anions, or recently even simultaneously. Concentrations of ions in a sample can differ for several orders of magnitude and ions which are targeted as analytes also define a matrix for the determinations of other ions. The requirement that major and minor ions in a sample should be determined simultaneously with ion chromatography as completely as possible adds an additional dimension to the considerations what is for a particular sample and under particular fitness for purpose requirements the limiting concentration of an ion that can still be successfully determined and how uncertain are concentrations of other ions determined above this limit.

This research is focused on the uncertainty sources and limitations in determining a total ionic composition of natural water with ion chromatography. The chapter is organised into sections corresponding to the steps performed during an analytical process. Each section is focused on a particular case study with the objective of recognising and evaluating uncertainty contributions with a bottom-up approach. Methodology necessary for these evaluations is described and applied to give answers on the importance of particular contributions. This chapter tends to develop guidance and strategies applicable to new complex cases.

Outline of content:

- Influence of impurities in chemicals on the uncertainty associated with a concentration of an ion in a stock combined standard solution

A case study is based on some selected central European brines of different types and provides an insight into influence that impurities in chemicals can have on the uncertainties associated with the concentrations of ions in a combined stock standard solution.

- Uncertainty associated with a concentration of an ion introduced into a final multi-ion calibration solution as a separate standard and combined there with a stock combined standard solution containing the same ion as an impurity

http://dx.doi.org/10.5772/59012

A case study is focused on magnesium ion introduced into the final set of calibration standards as a standard solution and combined there with a multi-ion stock standard solution containing other ions of interest and magnesium ion as an impurity originating from other chemicals.

- A Horwitz-like uncertainty function derived on a model of seawater analysed with ion chromatography

The uncertainty functions known by now either tend to be entirely general as it is the case with the Horwitz function [15, 16] or in contrast to this focus on a particular analyte determined in a particular matrix with a particular method. On the contrary this research is sample focused. We evaluate if an ion independent correlation can be recognised if relative fractions of ions in the total ionic composition of samples are related to the relative uncertainties in determining their concentrations with interpolation from calibration line equations. The results of ion chromatographic analyses of seawaters comprising 156 data serve as an appropriate model because of wide concentration ranges of ions in a sample.

- An uncertainty prediction tool for determining simultaneously minor and major ions with ion chromatography

For two different ways of expressing relative fractions of ions in a sample the meaningful ion independent relations in a form of the Horwitz-like uncertainty functions can be confirmed. The prediction tools established through this research enable recognition of ions for which it can be expected that they are determinable simultaneously with ion chromatography with acceptable uncertainties. By setting appropriate criteria the prediction tools can be accustomed to the user defined requirements.

2. Influence of impurities in chemicals on the uncertainty associated with a concentration of an ion in a stock combined standard solution

Among analytical methods relative methods which require calibration highly prevail, and ion chromatography is a relative method. A calibration procedure usually starts with a stock standard calibration solution from which a set of calibration standards with different concentrations is prepared. In techniques which enable a simultaneous determination of several analytes, as it is a case with ion chromatography, a combined standard solution comprising several ions is required. Some multi-ion stock standard calibration solutions which have concentrations of ions certified together with their combined standard uncertainties or expanded uncertainties are commercially available. Unfortunately they are of a limited usability when it comes to the analyses of natural waters with higher mineralisation. The concentrations of ions in commercial standards are usually far too similar to be applicable to analyses of waters in which concentrations of ions can extend over several orders of magnitude. Consequently a stock combined standard solution should be prepared from standard substances in a laboratory in order to better reflect the composition of samples which one tends to analyse.

The most desirable in instrumental analyses are linear calibration functions. The ordinary least squares regression model (OLS) is most frequently used for fitting the straight line through calibration points. One of the requirements for applying the OLS regression model is that concentrations of calibration solutions bear a negligible standard error if compared with the standard error of a measured signal. Consequently the uncertainty in concentrations of calibration solutions is not considered in estimating a standard uncertainty of interpolation of a concentration of an ion in a sample from a calibration line equation. But anyway a concentration of a stock combined standard solution already is associated with a certain standard error or uncertainty which should be considered and taken into account. As it was already mentioned in introduction for cases similar to this when a particular uncertainty contribution does not reflect directly in a model equation, it is taken into account at the end of a process and combined with the standard uncertainty estimated previously from the model equation. The law on propagation of errors for the operations of summation is applied.

Uncertainty of a concentration of a single component stock calibration standard solution depends on a procedure and quality of equipment used in a preparation of a solution, and also on purity of a chemical. The most usual way of preparing a stock combined standard solution is by weighing. An adequate mass of a chemical is weighed on an analytical balance; the chemical is transferred quantitatively into a volumetric flask, dissolved and diluted with a solvent to a final volume. Evaluation of uncertainty of a single component stock calibration solution concentration in such a case is straight forward taking into account the three main uncertainty contributions, namely uncertainty in measuring mass and volume and uncertainty associated with the purity of a chemical. Several easy to follow examples of such evaluations can be found in literature [7, 17].

But when it comes to stock combined standard solutions with concentrations of components extending over a wide concentration range a situation is more complicated. In such a case the presence of other chemicals in the same solution brings in another uncertainty contribution originating from the impurities present in other chemicals. We developed methodology for evaluating uncertainties associated with mass concentrations of ions in stock combined standard solution and proved that ion originating as an impurity from other chemicals can represent an important contribution to the combined standard uncertainty associated with the concentration of this ion primarily originating from its mother salt; and that this aspect was previously overlooked as we pointed out [18]. Here we apply this methodology to a case study relevant for water quality data evaluation with an objective of learning lessons on the importance of particular uncertainty contributions under different circumstances and with the objective of refining and simplifying the methodology for a further intended use.

2.1. A case study outline

A case study focuses on real central European brines of different types with different concentration proportions of chloride and sulfate. We intentionally limit ourselves on only two ions because in practice it is less likely that an additional ion has an impact as important as the previous one already influencing the uncertainty budget of the primary ion. We examine the influence of the presence of another chemical in a combined standard solution on the uncertainty associated with a concentration of the ion originating from a chemical of its prime origin and consider different purities of a prime chemical and different mass fractions of the ion present as an impurity in another chemical, and give answers on the importance of particular uncertainty contributions.

2.2. A choice of water samples for a study and characteristics of water samples

Data on a composition of natural waters, which we use in this study, originate from an extensive compilation of water analyses of the Central European mineral and thermal waters, which was prepared by Carlé [19]. In order to make these data manageable and to be able to exploit their full potential we prepared an electronic database comprising all essential information on more than 700 water samples. From this database we selected ten water samples with different proportions of mass concentrations of sulfate and chloride ions. The requirement was to include samples with a proportion of the two ions extending over a range as wide as possible but at the same time not allowing that mass concentration of any of the two ions becomes so low that uncertainty in weighing the chemical from which the ion originates becomes an essential influential factor in the uncertainty budget associated with the concentration of this ion. We ensured that the uncertainty in weighing a chemical was in the worst case below 40 % of the magnitude of the contribution of the most influential factor. Another requirement which we set was that at least one of the two ions determines a water type, meaning that it represents a major component.

A water type describes a water sample in terms of the most prevailing ions and can indicate to some extent a water genesis or a hydrogeological origin. A water type is determined by naming the major ions in a sequence corresponding to their decreasing concentration. The cations are named before the anions. An ion qualifies to be named in a water type by at least reaching by its concentration a limit of 20 milliequivalent percent. A judgement in this non SI unit is made possible by converting the results of an analysis from mass concentrations, usually expressed in mg/L, into milliequivalent (meq) concentrations. Mass concentration of an ion is divided by its molar mass and multiplied by the charge number of the ion under consideration. Anions and cations are treated separately. A total milliequivalent concentration of cations is calculated by a summation of milliequivalent concentrations of all catons. A milliequivalent percent of a cation is calculated as a fraction of the total milliequivalent concentration and expressed in percent. The same procedure is applied to anions. An example is a Ca-Mg-HCO3-SO4 type of water; or in words calcium, magnesium, hydrogen carbonate, sulfate type. The presence of dissolved CO_2 can also be indicated in a water type. The criterion which applies in such cases is that CO_2 concentration should at least reach a limit 200 mg/L. An example is a water of a Na-HCO3-CO2 type.

The waters, which qualified for a study, are summarised in table 1. They are identified by their place and state of origin and classified by a water type and further defined by mass concentrations of chloride and sulfate ions. Mass concentrations of ions under consideration are indicated in bold. Another ion defines a matrix. A concentration proportion of the matrix ion and ion under consideration is also given and marked in bold, either sulfate and chloride ratio or chloride and sulfate ratio.

Place	State	Water type	$\gamma(Cl^-)$ / mgL^{-1}	$\gamma(SO_4^{2-})$ / mgL^{-1}	SO_4^{2-}/ Cl^-	Cl^-/ SO_4^{2-}
Budapest	Hungary	Mg-Na-SO4	**49.7**	11982.9	**241.1046**	0.004148
Forste	Deutschland	Na-Cl-SO4	**36.5**	4360.4	**119.463**	0.008371
Nateln	Deutschland	Na-Cl-CO2	**22.4**	1589	**70.9375**	0.014097
Leer	Deutschland	Na-Ca-Cl	**580**	27728.16	**47.80717**	0.020917
Laa an der Thaya	Austria	Mg-Ca-SO4	**2515.9**	12270.2	**4.877062**	0.205041
Grunewald	Deutschland	Na-Cl	4204	**1577**	0.375119	**2.665821**
Buggingen	Deutschland	Na-Cl	50200	**2400**	0.047809	**20.91667**
Purbach	Austria	Mg-Na-SO4-Cl	197426.5	**1016.4**	0.005148	**194.2409**
Winzlar	Deutschland	Ca-Mg-SO4	153727	**335.7**	0.002184	**457.9297**
Bad Niedernau	Deutschland	Mg-Na-SO4-CO2	145386	**202**	0.001389	**719.7327**

Table 1. Characteristics of water samples selected for a study [19]

A water type gives a first impression on the composition of a water sample. A more comprehensive presentation of the characteristics of natural water is in a Piper diagram. Equally as for a water type the composition of a water sample has to be expressed in milliequivalent percent in order to be represented in a Piper Diagram. A Piper diagram has three sections; a triangular one on the left for a representation of a cationic composition, a triangular section on the right for an anionic composition, and a diamond on the top to represent a water sample as a whole in terms of anionic and cationic composition. A pair of dots in the two triangles of which one represents the cationic composition and another the anionic composition of the same water sample is transferred into a diamond section by drawing two lines and finding the intersection between them and representing the intersection by a dot. The procedure is that each line is drawn though one of the points parallel to the external edge of a triangle to which the dot pertains. A Piper diagram demonstrates that the dots distribute nearly evenly along the whole extent of the left upper edge of a diamond. Waters which position themselves into left corner of a diamond are brines and those in the upper corner of a diamond are characterised by high permanent water hardness. A distribution of the dots in a diagram confirms that the choice of the ten water samples was reasonable also in terms of the diversity in their composition.

The data summarised in table 1 represent models for preparations of stock combined calibration standard solutions. The lowest concentration of an ion under consideration is 22.4 mg/L and the highest 2515.9 g/L. The lowest concentration of a matrix ion is 1.589 g/L and the highest 197.4 g/L. Nine different water types are included. Proportions of mass concentrations of the matrix ion and the ion under consideration extend over three orders of magnitude. These water samples are so different that for a preparation of calibration solutions with appropriate concentrations of the two ions each of the samples would require its own stock combined calibration standard solution similar to it in its composition. By a choice of these ten water samples there are two sets of five uncertainty cases to be considered. In the first five cases,

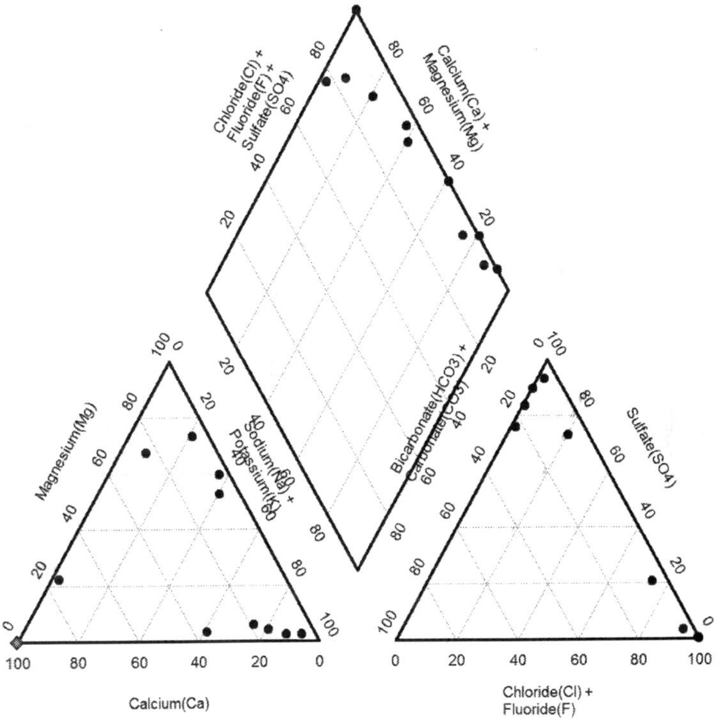

Figure 1. Ionic composition of ten central European water samples selected for evaluation of the influence of impurities in chemicals on the uncertainty associated with a concentration of an ion in a stock combined calibration standard

which are defined by a mass ration of sufate and chloride ion extending between 241 and 4.88, the chloride ion present as an impurity in a sodium sulphate chemical contributes to the uncertainty associated with the chloride concentration primarily originating from a sodium chloride chemical. In the second five cases, which are defined by a mass ratio of chloride and sufate ion between 2.67 and 720, the sulfate ion present as an impurity in a sodium chloride chemical contributes to the uncertainty associated with the sulfate concentration primarily originating from a sodium sulfate chemical.

2.3. A choice of chemicals for a study

For evaluating uncertainty associated with a concentration of a stock combined calibration standard solution apart from selecting the composition of model solutions in terms of their chloride and sulfate concentrations other parameters had to be defined. A decision was that a class A, 500 mL volumetric flask, provides a good compromise between a volume of a solution and masses of chemicals required for a preparation of a stock combined calibration standard solutions. The lowest mass of a chemical weighed under these circumstances is 18.5 mg of sodium chloride required for a preparation of a solution, which models a composition of water from Nateln, Deutschland.

The data on a quality of commercial chemicals, sodium chloride and sodium sulfate were collected. The specifications on their purity and presence of another ion as an impurity are summarised in table 2. A code for a level of purity and level of impurity was assigned to each chemical for easy marking. H stands for higher and L for lower level of purity. The level of impurity is graded with numbers from one to six, one indicating the lowest level of impurity, and six the highest. The symbol w_{salt} stands for a mass fraction of a salt e.g. NaCl or Na_2SO_4, which is assigned to a chemical on the basis of a specification given by a producer. This value is used in calculations and uncertainty evaluations. It is calculated as a middle of a range between the lowest expected purity specified by a producer and the highest possible purity, 100 %, and divided by 100 so that it is expressed as a fraction of one, instead of percent. A mass fraction of an impurity indicated by w_{impurity} is obtained similarly as a middle of the range, but in this case the specified value represents the highest expected impurity and the lowest possible is 0 %.

Chemical	NaCl	NaCl	Na_2SO_4	NaCl	Na_2SO_4	Na_2SO_4	NaCl	NaCl
Specified purity (%)	>99.5	>99.5	>99.5	>99.5	>99.0	>99.0	>99.0	>99.0
Purity code	H	H	H	H	L	L	L	L
w_{salt}	0.9975	0.9975	0.9975	0.9975	0.995	0.995	0.995	0.995
Impurity	SO_4^{2-}	SO_4^{2-}	Cl^-	SO_4^{2-}	Cl^-	Cl^-	SO_4^{2-}	SO_4^{2-}
Specified impurity (%)	<0,001	<0,01	<0.02	<0.05	<0.0005	<0.001	<0.004	<0.02
Impurity code	2	4	5	6	1	2	3	5
w_{impurity}	0.000005	0.00005	0.0001	0.00025	0.0000025	0.000005	0.00002	0.0001

Table 2. The characteristics of chemicals selected for a study

Eight combinations of sodium chloride and sodium sulfate chemicals were selected for a study. The combinations were generalised irrespective of a role of a particular chemical and coded in accordance with the purity level of prime chemical and impurity level of the ion under consideration in the secondary chemical. The codes were: L6, L2s, L2c, L1, H1, H2, H4 and H5. The first two and the seventh combination are dedicated to the consideration of the uncertainty associated with the concentration of sulfate ion. All others pertain to the considerations of chloride ion concentration and its uncertainty. In the first four cases the chemicals which are the prime source of an ion are of a lower purity, L. In the remaining four cases the chemicals were of a higher purity, H. A number as indicated in table 2 gives information on the impurity level in the secondary chemical. A stock combined standard solution corresponding to H5 and dedicated to a consideration of uncertainty associated with the chloride ion concentration could have been prepared from any sodium chloride chemical with purity higher than 99.5 % and the sodium sulfate chemical containing chloride ions in a mass fraction lower than 0.02

%. The combinations L2 are indistinguishable in terms of the quality of the two chemicals used for a preparation of each solution. The difference is in the ion under consideration. Hence an additional letter is assigned to these two combinations; a letter s stands for sulfate and c for chloride.

2.4. Uncertainty evaluation and uncertainty budget

Uncertainty evaluations were performed with a spreadsheet method. Each combination of chemicals which is dedicated to evaluation of the uncertainty associated with a concentration of chloride ion in a stock combined standard solution, namely L2c, L1, H1, H2 and H5 was considered for a preparation of five stock combined standard solutions, corresponding in their composition to first five examples of water samples listed in table 1. The remaining five water compositions were considered in relation to the sulfate ion uncertainty budget and combinations of chemicals L6, L2s and H4. Consequently 40 cases were evaluated altogether.

An uncertainty evaluation starts with an analysis of a process or procedure, which is in this case simple enough and does not require a presentation in a form of a flowchart. The two chemicals are weighed and transferred quantitatively into a 500 mL, class A volumetric flask. The solution is diluted with purified water to its final volume.

Model equation 1 which is commonly used in literature for calculation of a mass concentration obtained by a preparation of a solution by weighing takes into account only a direct source of an ion in a chemical of its prime origin. The symbols γ_{ion}, $m_{ion\_due\_to\_source}$ and $V_{solution}$ stand for a mass concentration and mass of the ion in a solution and volume of a solution, respectively. Mass of a salt weighed for a preparation of a solution is indicated by m_{salt}. The symbols M_{ion} and M_{salt} stand for molar mass of an ion or salt, respectively. It is necessary mentioning that in all the cases considered here a stoichiometric relation between an ion and the salt of its origin is one. If this is not the case a stoichiometric factor has to be introduced into equation.

$$\gamma_{ion} = \frac{m_{ion\_due\_to\_source}}{V_{solution}} = \frac{m_{salt} \cdot w_{salt} \cdot M_{ion}}{M_{salt} \cdot V_{solution}} \tag{1}$$

If other chemicals which contain the same ion as an impurity are present in a solution in concentrations higher than the ion under consideration additional parameters should be included and taken into account (18). Consequently equation 1 should be rewritten to give equation 2.

$$\gamma_{ion} = \frac{m_{ion\_due\_to\_source} + m_{ion\_from\_another\_chemical}}{V_{solution}} = \frac{\frac{m_{salt} \cdot w_{salt} \cdot M_{ion}}{M_{salt}} + m_{other\_salt} \cdot w_{impurity}}{V_{solution}} \tag{2}$$

A cause and effect diagram indicating all uncertainty sources associated with a preparation of a stock combined standard solution and pertaining to equation 2 is presented in figure 2 (left) and compared to a diagram usually used (right). What distinguishes the diagram on the left from the one usually drawn is a branch in its lower part indicated with $m_{\text{ion\_from\_another\_chemical}}$ and corresponding to a mass of an ion originating as an impurity from another chemical.

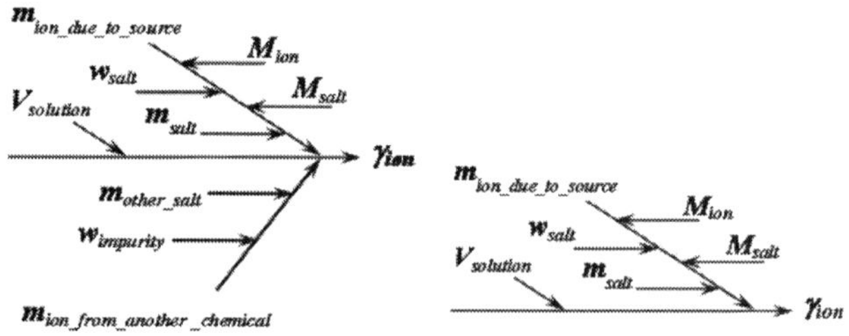

Figure 2. A cause and effect diagrams identifying contributions to the uncertainty associated with a mass concentration of an ion in a stock combined calibration standard; left pertaining to equation 2; right pertaining to equation 1.

In the continuation we outline the main steps undertaken during the uncertainty evaluation, more details can be found elsewhere [8]. An uncertainty statement relating to equation 2 is given by equation 3.

$$u(\gamma_{\text{ion}}) = \frac{m_{\text{ion\_due\_to\_source}} + m_{\text{ion\_from\_another\_chemical}}}{V_{\text{solution}}} \sqrt{\left(\frac{u\left(m_{\text{ion\_due\_to\_source}} + m_{\text{ion\_from\_another\_chemical}}\right)}{m_{\text{ion\_due\_to\_source}} + m_{\text{ion\_from\_another\_chemical}}}\right)^2 + \left(\frac{u(V_{\text{solution}})}{V_{\text{solution}}}\right)^2} \quad (3)$$

Equation 4 combines uncertainty sources contributing to the uncertainty associated with a mass of an ion originating directly from a chemical of its prime origin.

$$u(m_{\text{ion\_due\_to\_source}}) = m_{\text{ion\_due\_to\_source}} \sqrt{\left(\frac{u(m_{\text{salt}})}{m_{\text{salt}}}\right)^2 + \left(\frac{u(M_{\text{salt}})}{M_{\text{salt}}}\right)^2 + \left(\frac{u(M_{\text{ion}})}{M_{\text{ion}}}\right)^2 + \left(\frac{u(w_{\text{salt}})}{w_{\text{salt}}}\right)^2} \quad (4)$$

For the quantification of the measurement uncertainty associated with weighing by difference, equation 5 derived from the calibration certificate following the DKD-R 7-1 calibration procedure was used:

$$u(m_{\text{salt}}) = 0.0105 \cdot \text{mg} + 1.7 \cdot 10^{-6} m_{\text{salt}} \quad (5)$$

The standard uncertainty associated with the purity of a chemical is derived from a half of the purity range by taking into account appropriate probability function. A half of the purity range is defined by a difference between 1 corresponding to 100 % purity, and w_{salt}, which is a mass fraction of the salt in a chemical which corresponds to the middle of the purity range specified by a producer. A rectangular distribution is most frequently used assuming that all values within a specified range are of equal probability. The rectangular probability function is taken into account by dividing a half of the range by a square root of three. We used this approach in equation 6, even though some authors claim that different probability functions should be applied to the purities determined by different analytical methods, recognising that the rectangular distribution is only fit for purpose if purities were determined by titrimetry. If chromatographic methods were used a triangular-ramp probability function is more relevant [20]. We did not distinguish these two cases due to lack of information on the testing methods used by a producer of a chemical.

$$u(w_{\text{salt}}) = \frac{1 - w_{\text{salt}}}{\sqrt{3}} \tag{6}$$

Uncertainty associated with a mass of an ion originating from another chemical has two contributions, uncertainty in measuring a mass of this chemical indicated by $m_{\text{other\_salt}}$ and uncertainty in the mass fraction of the impurity in a chemical. As equation 7 demonstrates the second term under a square root simplifies into an inversion of a square root of three. The reason is that w_{impurity} indicates two parameters which are numericaly equal. Firstly it indicates the middle of the impurity interval defined by the maximal expected impurity level declared by a producer and the lowest theoretically possible, zero impurity. Secondly it also represents the half width of this interval, which is transformed into a standard uncertainty by dividing by a square route of three.

$$u(m_{\text{ion\_from\_another\_chemical}}) = m_{\text{other\_salt}} \cdot w_{\text{impurity}} \sqrt{\left(\frac{u(m_{\text{other\_salt}})}{m_{\text{other\_salt}}}\right)^2 + \left(\frac{w_{\text{impurity}}}{w_{\text{impurity}}\sqrt{3}}\right)^2} \tag{7}$$

If it is proved that uncertainty contribution associated with the weighing is negligible if compared with the impurity contribution, equation 7 can be simplified into equation 8.

$$u(m_{\text{ion\_from\_another\_chemical}}) = \frac{m_{other\_salt} \cdot w_{impurity}}{\sqrt{3}} \tag{8}$$

Uncertainty associate with the volume measurement is estimated as specified by equation 9. The symbols $V_{\text{toler\_as\_}\pm}$, $u(V_{\text{rep}})$ and $T_{\text{variability\_as\_}\pm}$ stand for tolerance of a calibration volume, uncertainty due repeatability of filing a volumetric flask and expected temperature variations around the temperature of calibration.

$$u(V_{\text{solution}}) = \sqrt{\left(\frac{V_{\text{toler\_as\_}\pm}}{\sqrt{6}}\right)^2 + \left(u(V_{\text{rep}})\right)^2 + \left(\frac{V_{\text{soluton}} \cdot 2.1 \cdot 10^{-4} \cdot {}^\circ\text{C}^{-1} \cdot T_{\text{variability\_as\_}\pm}}{\sqrt{3}}\right)^2} \tag{9}$$

Uncertainty evaluations were performed with a spreadsheet method in order to gain understanding on the importance of the particular uncertainty contributions under different circumstances. For each of 40 cases a bar diagram on uncertainty contributions was drawn. We are not going to give a detailed account of the spreadsheet method which we used at this stage, due to a fact that some contributions as we explain later can be omitted and a model simplified. Four examples of the results of uncertainty evaluations are presented in figures 3 and 4 in a form of bar graphs. The bar graphs in figure 3 correspond to the evaluation of the uncertainty associated with a chloride ion concentration. The graph on the left was obtained for the L2c case evaluated at a sulfate-chloride mass ratio 4.88. The graph on the right relates to the H1 case evaluated at a sulfate-chloride mass ratio 70.9. The bar graphs in figure 4 correspond to the uncertainty evaluation associated with the sulfate ion concentration, the L6 and H4 cases and chloride-sulfate mass fractions 2.67 and 194 apply, respectively. Full details are given in text accompanying the figures. The expression $u(\gamma)/\gamma$ indicates a relative combined standard uncertainty in mass concentration of an ion. Text on the ordinate corresponds to $u(w_{\text{impurity}})$, $u(m_{\text{other\_salt}})$, $u(V_{\text{solution}})$, $u(M_{\text{salt}})$, $u(M_{\text{ion}})$, $u(w_{\text{salt}})$, $u(m_{\text{salt}})$ and $u(\gamma)$ if read from top to bottom.

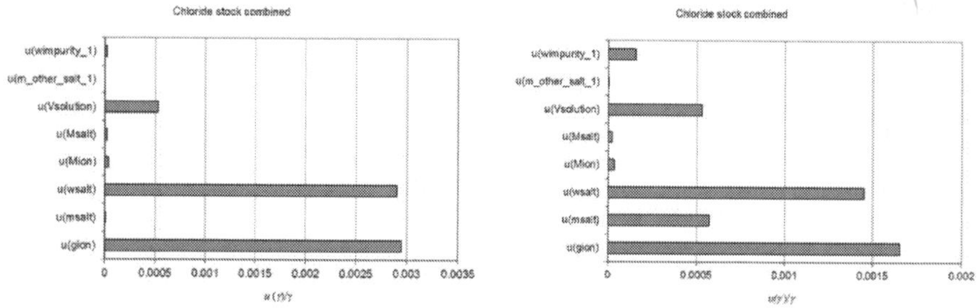

Figure 3. Uncertainty contributions associate with relative standard uncertainty of chloride concentration in a stock combined standard solution. The L2c example on the left: a sulfate-chloride mass ratio 4.88, sodium chloride purity 99.5 % and chloride impurity in sodium sulfate 0.001 %. Composition corresponds to a water of a Mg-Ca-SO4 type from Laa an der Thaya, Austria. The H1 example on the right: a sulfate-chloride mass ratio 70.9, sodium chloride purity 99.75 % and chloride impurity in sodium sulfate 0.0005 %. Composition corresponds to a water of a Na-Cl-CO2 type from Nateln, Deutschland.

A similarity between the two cases in figure 3 is that the impurity levels, which are indicated by a number in a code, are the lowest two of all. Far the greatest contribution to the combined standard uncertainty in both cases is due to uncertainty associated with the purity of a prime chemical. Consequently the relative combined standard uncertainty is in the first case slightly below 0.003 or 0.3 % and in the second case close to 0.0017 or 0.17 %. The distinction between these two cases is that uncertainty in measuring a mass of the prime chemical contributes to the combined standard uncertainty in the second case even to a slightly higher extent than uncertainty in a volume measurement which is similar in both cases. The reason is that a mass of sodium chloride measured for a preparation of the stock combined standard solution that corresponds to a composition of a water from Nateln, Deutschland was the lowest of all, only

http://dx.doi.org/10.5772/59012

18.5 mg. Even though the impurity level was the lowest of all in the H1 example and a sulfate-chloride mass ratio with its value 70.9 was moderate, a mass fraction of impurity in another chemical starts to indicate a contribution.

The L6 example in figure 4, left, which corresponds to the highest impurity level, indicates that uncertainty in a mass fraction of the impurity in another chemical contributes to the standard combined uncertainty already as the second major factor exceeding the uncertainty contribution associated with a volume measurement, even though the chloride-sulfate mass ratio is by 2.67 the lowest of all examined. A bar graph on the right side of figure 4, which corresponds to the H4 example, shows a graphical pattern that differs from all previous cases. In spite of the fact that the impurity level is only third the highest and the chloride-sulfate mass fraction is by 194 far away from the highest examine 720; uncertainty associated with the mass fraction of the impurity in a chemical is the major factor contributing to the relative combined standard uncertainty associated with the concentration of sulfate ion in a solution. The influence of the uncertainty in a purity of a prime chemical is small and uncertainty in volume measurement becomes negligible under such circumstances.

These four examples demonstrate that different combinations of experimental conditions lead to uncertainty budgets with very different profiles. In the continuation we are going to focus on the lessons that can be learned from all the 40 uncertainty evaluations performed.

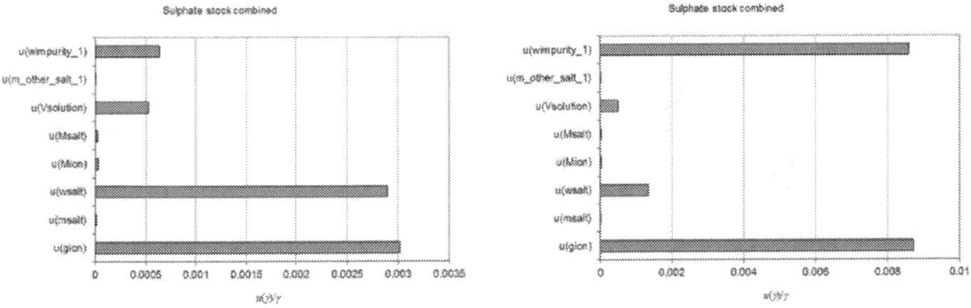

Figure 4. Uncertainty contributions associated with relative standard uncertainty of sulfate concentration in a stock combined standard solution. The L6 example on the left: chloride-sulfate mass ratio 2.67, sodium sulfate purity 99.5 % and sulfate impurity in sodium chloride 0.05 %. Composition corresponds to a water of a Na-Cl type from Grunewald, Deutschland. The H4 example on the right: chloride-sulfate mass ratio 194, sodium sulfate purity 99.75 % and sulfate impurity in sodium chloride 0.01 %. Composition corresponds to a water of a Mg-Na-SO4-Cl type from Purbach, Austria.

2.5. Lessons learned

Figure 5 demonstrates the influence that a mass ratio of the matrix ion and the ion under consideration has on the combined relative standard uncertainty associated with a mass concentration of the ion under consideration. Different combinations of a mass fraction of a salt in a prime chemical and a mass fraction of the ion under consideration present as an impurity in another chemical were considered. Each curve pertains to a particular combination of the source and matrix ion chemicals indicated by a code in a legend. Different scales are used in graphs on the left and right.

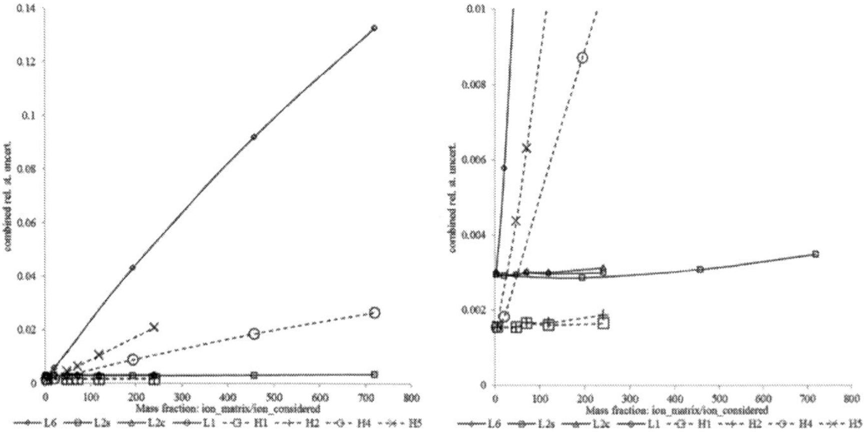

Figure 5. Influence of a mass ratio of a matrix ion and an ion of interest on a relative standard uncertainty associated with a concentration of the ion of interest at different purities of the ion source chemical and different impurity levels of this ion in another chemical.

Figure 6 gives an insight into influence that uncertainty associated with a mass fraction of the ion present as an impurity in another chemical has on the final combined standard uncertainty associated with the concentration of this ion in a stock combined standard solution. The results of the uncertainty estimations are expressed relatively on the ordinate axis. A contribution of the uncertainty associated with a mass fraction of the ion present as an impurity in another salt to the final combined standard uncertainty was divided with the final combined standard uncertainty. The meaning of abscise is the same as in figure 5. The curves for all eight combinations of the source and matrix ion chemicals which were considered are presented. The codes of the combinations are given in a legend. As figure 6 demonstrates if the ion under consideration is present in another salt as an impurity in a mass fraction not exceeding 0.01 % or 0.02 % or 0.05 % and a mass ratio of the matrix ion and an ion under consideration is higher than 200 than the uncertainty associated with a mass fraction of the impurity contributes by being close to 1 nearly 100 % to the standard combined uncertainty associated with a mass concentration of the ion.

Lessons learned from figures 5 and 6:

- The uncertainty budget of the stock combined standard solutions, which imitate the composition of natural waters, changes significantly with the change in the mass ratio of the matrix ion and ion under consideration and in relation to the mass fraction of a salt in a prime chemical.

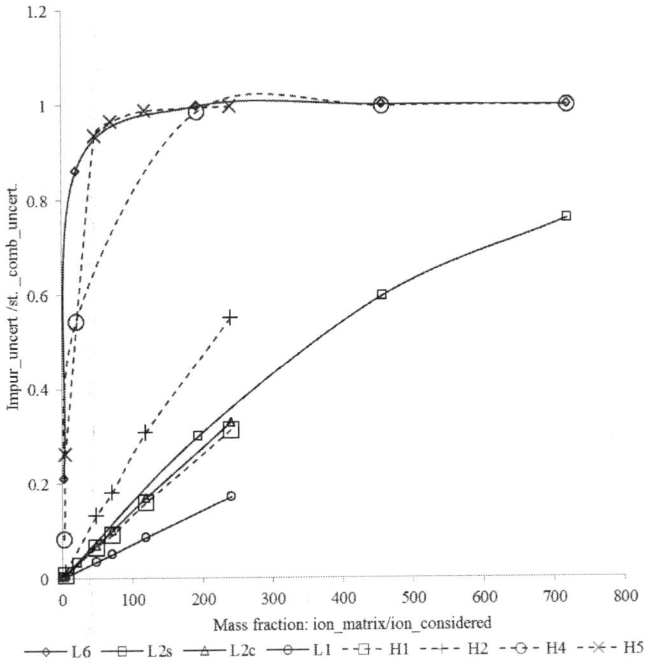

Figure 6. Influence of a mass ratio of a matrix ion and an ion of interest on a relative contribution of the uncertainty associated with the impurity mass fraction to the combined standard uncertainty in concentration of the ion of interest at different purities of the ion source chemical and different impurity levels of this ion in another chemical

- Mass fraction of an ion present as an impurity in another chemical should be taken into account in uncertainty evaluations; otherwise the combined standard uncertainty associated with the concentration of an ion in a stock combined standard solution can be underestimated.
- A mass fraction of a salt in a prime chemical is a decisive factor determining the relative combined standard uncertainty at low mass ratios of the matrix ion and ion under consideration.
- If a purity of the prime chemical is not lower than 99.0 % and an ion under consideration is present in another chemical as an impurity in a mass fraction lower than 0.001 % a relative combined standard uncertainty lower than 0.4 % can be expected for a mass concentration of an ion in a combined standard solution even in the most unfavourable cases in which another ion strongly prevails, as it is the case with sulfate ion in a solution with a chloride-sulfate mass ratio 720. At lower matrix ion and ion under consideration ratios relative combined standard uncertainties around 0.3 % can be expected.

- If a prime chemical is of a higher, at least 99.5 % purity and an ion under consideration is present in another chemical as an impurity in a mass fraction lower than 0.001 % a relative combined standard uncertainty lower than 0.2 % can be expected for mass ratios of the matrix ion and ion under consideration lower than 250.
- At higher mass fractions of impurities than previously stated the relative combined standard uncertainty starts to rise rapidly with a mass ratio of the matrix ion and ion under consideration.
- With a mass fraction of the impurity in a secondary chemical up to 0.01 % or 0.02 % or 0.05 % a relative standard combined uncertainty associated with a mass concentration of the ion under consideration remains below 0.4 % only if a mass ratio of the matrix ion and ion under consideration does not exceed 80 or 40 or 10, respectively.
- The relative standard uncertainty associated with the concentration of an ion in a stock combined standard solution is hardly acceptable if its magnitude reaches 1 %. With a mass fraction of the impurity in a secondary chemical not exceeding 0.01 % or 0.02 % or 0.05 % a relative standard combined uncertainty associated with the mass concentration of the ion under consideration exceeds 1 % if the mass ratios of the matrix ion and an ion under consideration exceeds 220 or 120 or 40, respectively.
- With a mass fraction of the impurity in a secondary chemical up to 0.05 % a relative standard combined uncertainty associated with the mass concentration of the ion under consideration can exceed 13 % in the most unfavourable cases in which another ion strongly prevails, as it is the case with sulfate ion in a solution with a chloride-sulfate mass ratio 720.
- If any of the uncertainty contributions in the uncertainty budget exceeds 30 % of the combined standard uncertainty it cannot be considered negligible. Impurity of the secondary chemical has larger effect on the uncertainty budget associated with the concentration of an ion in a stock combined standard solution if the prime chemical is of higher purity.
- If the mass ratio of the matrix ion and ion under consideration exceeds 240 and the purity of the prime chemical exceeds 99.5 % it can even be expected for the secondary chemical with the lowest mass fraction of the ion present as an impurity, lower than 0.0005 %, that the impurity contribution exceeds 30 % of the combined standard uncertainty in the uncertainty budget. With a chemical with a mass fraction of an impurity not exceeding 0.001 % a 30 % contribution to the uncertainty budget is reached at the mass ratio 120. With a chemical with a mass fraction of an impurity not exceeding 0.01 % this happens already at the mass ratio 5.
- With a mass fraction of the impurity in a secondary chemical up to 0.01 % or 0.02 % or 0.05 % the uncertainty associated with the impurity of a chemical exceeds 80 % of the combined standard uncertainty associated with the mass concentration of an ion in a stock combined standard solution if the mass ratio of the matrix ion and ion under consideration exceeds 10 or 45 or 100, respectively.

2.6. Refined methodology for uncertainty evaluations in prospect complex cases

Critical evaluation of all 40 bar graphs lead to a conclusion that only uncertainties associated with the mass fraction of an ion present as an impurity in another chemical, mass fraction of a salt in a prime chemical, mass of the prime chemical and volume of solution should really be taken into account. A refined cause and effect diagram which we suggest for prospect use is presented in figure 7, left. A mass of the prime chemical, although negligible in most cases, is not left out. When it comes to solutions with a high ration of matrix ion an ion under consideration a compromise between the volume of the solution not being too high and a mass of the prime chemical not being too low has to be found. This leads to masses which are below a limit recommended for a reliable weighing. Even though as solid bars in figure 7, right, demonstrate, the uncertainty contribution associated with weighing does not reach 40 % of the major uncertainty contribution and it is comparable in magnitude to the uncertainty contribution of the volume measurement, it cannot be considered non-existent. The bars with a pattern confirm that further simplification of a model not taking into account this contribution would have led to an underestimation of the combined standard uncertainty associated with a concentration of an ion in a stock combined standard solution.

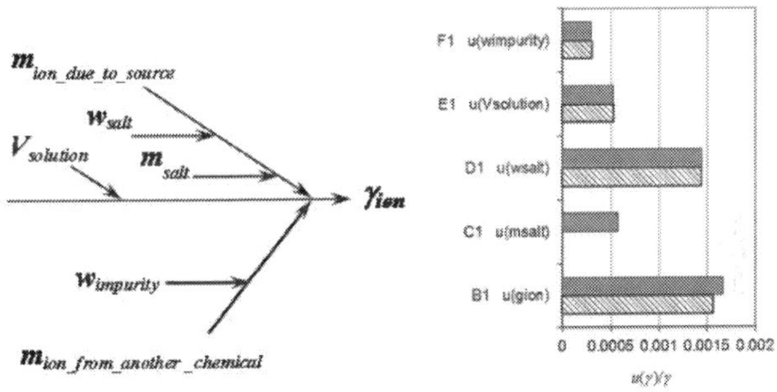

Figure 7. A refined cause and effect diagram, left, and uncertainty budget associated with it, right. Solid bars correspond to a cause and effect diagram on the left, bars with a pattern demonstrate the influence of a further simplification of a model by omitting the uncertainty associated with a mass of the prime chemical

A spreadsheet method which we suggest for evaluation of combined standard uncertainty in prospect cases is presented in table 3. Shaded cells are intended for input. Formulas in line 12 are not all fully visible but they are all similar to a formula in a cell B12, apart from a difference in a letter. Cells with frames indicate the main results of an uncertainty evaluation, namely a mass concentration of the ion under consideration in the stock combined standard solution, a standard uncertainty associated with it and a relative standard uncertainty pertaining to it. Data in line 16 or 17 can be used for a bar graph representation of the uncertainty contributions of particular parameters to the combined standard uncertainty. Numerical results are given below. Calculations were performed for H2 example and mass ratio 70.9 corresponding to a

A1	B1 u (γion)	C1 u (m salt)	D1 u (w salt)	E1 u (V solution)	F1 u (w impurity
		=0.0105+0.0000017*B4	=(1-B5)/(3^0.5)	0.26392738900440	=B10/3^0.5
A3 a ion/compound	1	1	1	1	1
A4 m salt (mg)	1165.90875381845	**=B4+C2**	=B4	=B4	=B4
A5 w salt	0.995	=B5	**=B5+D2**	=B5	=B5
A6 M ion (g/mol)	96.0626	=B6	=B6	=B6	=B6
A7 M salt (g/mol)	142.04213856	=B7	=B7	=B7	=B7
A8 V solution (mL)	500	=B8	=B8	**=B8+E2**	=B8
A9 m another_salt (mg)	3465.05799302062	=B9	=B9	=B9	=B9
A10 w impurity_1	0.00025	=B10	=B10	=B10	**=B10+F2**
A12 γ ion (g/L)	=(B3*B4*B5*B6/B7+B9*B10)/B8	=(C3*C4*C5*C6/C7+C9	=(D3*D4*D5*D6/	=(E3*E4*E5*E6/	=(F3*F4*F5*F6/
A13 u (γion,xi)		=C12-B12	=D12-B12	=E12-B12	=F12-B12
A14 u (γion)	=(C14+D14+E14+F14)^0.5	=C13^2	=D13^2	=E13^2	=F13^2
A16 ABS(u (γion,xi))	=B14	=ABS(C13)	=ABS(D13)	=ABS(E13)	=ABS(F13)
A17 u (γion,xi)/γion	=B16/$B$12	=C16/$B$12	=D16/$B$12	=E16/$B$12	=F16/$B$12
γ ion (g/L)	1.57084752899651				
ABS(u (γion,xi))	0.0047341372961131	0.0000167987106964862	0.00455240687256	0.00082874191832	0.0010002760
u (γion,xi)/γion	0.00301374717069922	0.0000106940427930759	0.00289805776087	0.00052757629434	0.0006367747

Table 3. A spreadsheet method suggested for prospect evaluations of uncertainty associated with a concentration of an ion in a stock combined standard solution

preparation of a stock combined standard solution simulating a composition of water from Nateln, Deutschland, table 1. The final result would have been written $\gamma(Cl^-) \pm U$=(1.571 ± 0.009) g/L; k=2.

3. Uncertainty associated with a concentration of an ion introduced into a final multi-ion calibration solution as a separate standard and combined there with a stock combined standard solution containing the same ion as an impurity

Another case, which requires consideration of the influence of impurities present in other chemicals on the uncertainty associated with a concentration of this ion appears when a stock combined standard solution has to be combined with a separate standard solution of this particular ion. Such cases materialise when a standard substance which would have allowed for a direct introduction of this particular ion into sock combined standard solution is not available and a commercial standard solution should be used instead, but is not available in a concentration high enough to be introduced into the stock combed standard solution directly, for a reason that a volume grater then a volume of a volumetric flask would have been required. Consequently this ion can be added only at the final stage when series of standard calibration solutions are prepared.

A case study focusses on magnesium ion introduced into the final set of calibration standards as a stock standard solution and combined there with a stock multi-ion combined standard

http://dx.doi.org/10.5772/59012

solution containing other ions of interest. Even though not added in a form of magnesium salt the magnesium ion is present in the combined stock standard solution as an impurity originating from other chemicals. Example is taken from the analyses of sea water, more precisely determination of cations with ion chromatography. Only the influence of magnesium ion present as an impurity in sodium chloride is considered for the reason that sodium is the only cation in sea water with a concentration higher than the concentration of magnesium ion. The sodium and magnesium mass ratio is 8.4.

A procedure carried out for a preparation of a final multi ion combined standard solution, and a cause and effect diagram pertaining to it, are presented in figure 8, left/right. A symbol γ_{ion} previously used in relation to the stock combined standard solution was replaced with $\gamma_{another\_sol}$ in order to enable a clear distinction of the two sources of magnesium ions originating from two solutions. Magnesium ions in a combined multi-ion calibration standard solution have two contributions $m_{ion\_st\_sol}$ originating from a magnesium ion commercial stock standard solution and $m_{another\_sol}$ originating from the stock combined standard solution. The symbols γ_{st_sol} and $\gamma_{another\_sol}$ and V_{st_sol} and $V_{another\_sol}$ stand for mass concentrations of magnesium ions in the two solutions and volumes of the two solutions measured with a piston micropipette. A mass concentration of magnesium ion in a final multi-ion calibration standard is indicated by $\gamma_{ion\_dil\_total}$ and a volume of this solution by V_{final}.

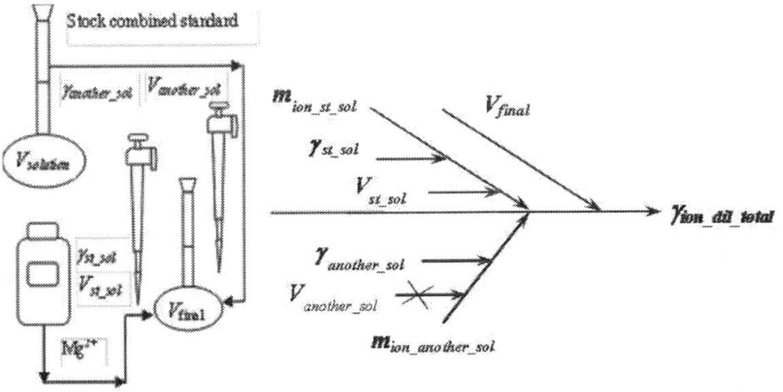

Figure 8. Influence of magnesium ion present as an impurity in a stock combined standard solution on the final concentration of magnesium ion and its uncertainty; a flow chart for a preparation of the final combined calibration solution and a cause and effect diagram, left/right

A model equation 10 defining a mass concentration of an ion in a solution prepared by a dilution of a stock standard solution, which is usually used and does not take into account the influence of ions present as impurities in another solution, has to be amended and rewritten to give a model equation 11.

$$\gamma_{ion\_dil\_total} = \frac{\gamma_{st\_sol} \cdot V_{st\_sol}}{V_{final}} \tag{10}$$

$$\gamma_{ion\_dil\_total} = \frac{\gamma_{st\_sol} \cdot V_{st\_sol} + \gamma_{another\_sol} \cdot V_{another\_sol}}{V_{final}} \tag{11}$$

Uncertainty contributions appearing in figure 8 and associated with volumetric operations are treated by following equation 9 but by omitting the last term associated with the deviation of the ambient temperature from the temperature of calibration. The reason is that all volumetric operations are performed during a short period of time, during which the ambient temperature is not expected to change, therefore it affects all the measurements proportionally.

Uncertainty associated with a mass concentration of magnesium ion in a commercial stock standard solution is declared by a producer as a range of a possible variation and transformed into standard uncertainty by dividing by a square root of three.

The only contribution requiring explanation is uncertainty associated with a mass concentration of magnesium ion in another solution, namely the stock combined standard solution containing sodium chloride. An expression for a mass concentration of an ion present as an impurity in another solution and originating from another salt is derived from equation 2 and reduced to equation 12 by taking into account that m_{salt} equals zero.

$$\gamma_{another\_sol} = \frac{w_{impur} \cdot m_{other\_salt}}{V_{solution}} \tag{12}$$

The uncertainty contribution associated with $\gamma_{\mathrm{another\_sol}}$ can be evaluated using a spreadsheet method presented in table 3 by substituting input data in cells B4 to B7 by zero, zero, one and one, respectively and providing relevant input data for cells B8 to B7, and E1, namely 100 mL, 2687.45 mg, 0.000005 and 0.0643402 mL for a preparation of a solution in 100 mL volumetric flask from 2687.45 mg of sodium chloride with magnesium ion impurity content lower than 0.001 %. This procedure gives results for $\gamma_{\mathrm{another\_sol}}$ and $u(\gamma_{\mathrm{another\_sol}})$ $1.3437 \cdot 10^{-04}$ and $7.758 \cdot 10^{-05}$ expressed in g/L, respectively. It turns out that the uncertainty contribution associated with the volume of a solution is negligible by being three orders of magnitude smaller than the contribution of w_{impur}. Consequently the uncertainty evaluation can be simplified as declared by equation 13. It was confirmed that the uncertainty estimation obtained by this equation is numerically equal to the one obtained by a spreadsheet method.

$$u(\gamma_{another\_sol}) = \frac{w_{impur} \cdot m_{other\_salt}}{V_{solution} \cdot \sqrt{3}} \tag{13}$$

A combined standard uncertainty associated with a concentration of magnesium ions in a final combined calibration standard solution can be evaluated by using a spreadsheet method as indicated in figure 9, left. In the example given here the final solution is prepared in a 20 mL volumetric flask by measuring 126 μL of the commercial magnesium stock standard solution with a concentration (1.001±0.002) g/L, and 100 μL of the stock combined standard solution containing magnesium ion as an impurity in concentration $1.3437 \cdot 10^{-04}$ g/L, as realised in the previous step.

Magnesium	u(γion)	u(γst_sol)	u(Vst_sol)	u(γanother_s	u(Vanother_	u(Vfinal)
		0.0011547	0.00048313	7.758E-05	0.0004113	1.70783E-05
γst_sol (g/L)	1.001	**1.0021547**	1.001	1.001	1.001	1.001
Vst_sol (mL)	0.126	0.126	**0.12648313**	0.126	0.126	0.126
γanother_sol (g/L)	0.00013437	0.00013437	0.00013437	**0.00021195**	0.00013437	0.000134373
Vanother_sol (L)	0.1	0.1	0.1	0.1	**0.1004113**	0.1
Vfinal (L)	0.02	0.02	0.02	0.02	0.02	**0.020017078**
γion_final_sol (mg/	6.30697186	6.31424648	6.33115227	6.30735976	6.30697463	6.30159084
u(γion_final_sol,xi)		0.00727461	0.02418041	0.0003879	2.7634E-06	-0.00538102
u(γion)	0.02582088	5.292E-05	0.00058469	1.5047E-07	7.6361E-12	2.89554E-05
ABS(u(γion_final_s	0.02582088	0.00727461	0.02418041	0.0003879	2.7634E-06	0.005381023
u(γion_final_sol)/γ	0.00409402	0.0011521	0.00381927	6.15E-05	4.3814E-07	0.000853915

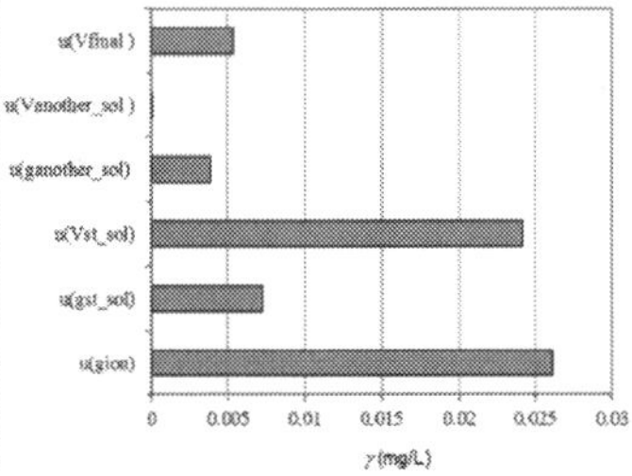

Figure 9. A spreadsheet method for evaluating the uncertainty associated with a concentration of an ion in a final combined calibration standard solution prepared by combining the ion stock standard solution and stock combined standard solution which contains the ion under consideration as an impurity (left), and the results obtained for magnesium ion and presented in a form of a bar graph (right)

As a bar graph in figure 9 demonstrates the uncertainty associated with the mass concentration of the magnesium ion present as an impurity in another solution, u(ganother_sol), contributes to the final combined standard uncertainty in spite of the fact that in the example which we considered the sodium ion mass concentration was only 8.4 times higher than magnesium ion concentration. This leads to a conclusion that in calculating the concentration and its uncertainty the influence of impurities should be taken into account when another ion is present in higher concentration than the ion under consideration and a multi-ion solution is prepared by combining the ion stock standard solution and the stock combined standard solution.

4. A Horwitz-like uncertainty function derived on a model of seawater

Total ionic composition of a sample can be determined with ion chromatography at least predominantly if not completely, and this can usually be achieved within two separate runs, one for cations and another for anions, or recently even simultaneously [3, 13, 21-23]. Concentrations of ions in a sample can differ for several orders of magnitude and ions which are targeted as analytes also define a matrix for the determinations of other ions. The requirement that major and minor ions in a sample should be determined simultaneously with ion chromatography as completely as possible adds an additional dimension to the considerations what is for a particular sample and under particular fitness for purpose requirements the limiting concentration of an ion that can still be successfully determined and how uncertain are concentrations of other ions determined above this limit.

Samples can be recognised as similar for ion chromatographic analyses not on the bases of concentration levels but on the bases of their relative proportions [24]. For the highest simultaneity of ion chromatographic determination of ions a sample dilution should be selected so that it prevents the major ions from overloading the column, and keeps their chromatographic peak areas linearly related to concentration, but at the same time lowers the concentrations of minor ions as little as possible. Hence sample dilutions reduce the number of distinctive analytical cases and set the ground for consideration if any general ion independent relation between the relative uncertainty of interpolation of concentrations from calibration lines, and fractions of ions in a sample can be recognised.

Every analytical result should be given together with the standard uncertainty assigned to it. Below a certain concentration limit the uncertainty of the result obtained with an analytical method becomes too high to be acceptable or tolerable and fit for purpose. Trying to determine the concentration at this limit is not feasible from the economical point of view. A prediction tool that could help making a correct consideration in advance could prevent such cases.

In the results of several independent inter-laboratory collaborative studies, in which different analytes in different matrices were targeted by different analytical techniques, William Horwitz recognised a general rule that the mean percent standard deviation $s_r(\%)$ as a measure of reproducibility of results, and concentration level of an analyte expressed relatively as a mass fraction (w), are related (equation 14). As follows from the equation each decrease in a concentration level of two orders of magnitude doubles $s_r(\%)$ [15, 25].

$$s_r(\%) = 2^{(1-0.5 \times \log w)} \tag{14}$$

The graphical representation of this relation is known as the Horwitz function and together with the Horwitz ratio [16] became a guidance for a judgement on the achievable and acceptable performance of an analytical method [26-28]. Horwitz estimated that a within laboratory $s_r(\%)$ ordinarily accounts for one-half to two-thirds of the total between-laboratory $s_r(\%)$, which comprises the within and between-laboratory variability contribution [15, 29].

Albert and Horwitz demonstrated that the Horwitz function can be rewritten in a slightly different form (equation 15) and assumed that the interlaboratory limit of quantitation corresponds to w of the order of magnitude 1×10^{-9} [30].

$$s_r(\%) = 2\ w^{-0.1505} \tag{15}$$

It was recognised that a Horwitz-like function holds for the results of proficiency tests [31, 32]. Thompson and Lowthian confirmed that experimental data follow the Horwitz function well down to w of 1×10^{-8} and that at the $s_r(\%)$ greater than 33 % the judgement on the presence of an analyte in the absence of other information is questionable [33].

Thompson later evaluated the results of more recent collaborative trials concerning analytes in sub 100 ppb concentration levels and observed greater deviations from the Horwitz function than previously observed, and suggested three different expressions for reproducibility standard deviation (s) for different concentration ranges (equation 16) [34].

$$s = \begin{cases} 0.22\ w & \text{if } w < 1.2 \times 10^{-7} \\ 0.02\ w^{0.8495} & \text{if } 1.2 \times 10^{-7} \leq w \leq 0.138 \\ 0.01\ w^{0.5} & \text{if } w > 0.138 \end{cases} \tag{16}$$

Linsinger and Josephs questioned the applicability of the Horwitz equation as a performance criterion for analytical methods [35]. Thompson objected the methodology they used and opposed a generalisation of their conclusions [36]. He suggested the uncertainty characteristic function as a tool for summarising or specifying the behaviour of an analytical system [37] and as a method-specific alternative to the Horwitz function [38] (equation 17). The relative standard deviation is denoted by s_r.

$$s=\sqrt{\alpha^2+(\beta\times w)^2}\ or\ s_r=\sqrt{\frac{\alpha^2}{w^2}+\beta^2} \tag{17}$$

In contrast to the Horwitz function the characteristic function follows the principles of error propagation theory and can therefore represent adequately the standard uncertainty of the final result obtained by an analytical method and apart from this also accommodates a detection limit, that can be derived from α. A form of the function is general but parameters α and β should be determined for each method separately through a method validation.

Côté et *al* applied the Horwitz equation and the uncertainty function suggested by Thompson to the data derived from the proficiency testing schemes aiming at the quantification of cadmium, lead and mercury in blood and urine and confirmed that the uncertainty function describes well the dependence of the reproducibility standard deviation on the mass fraction of an analyte [39]. In accordance with already stated it comes with no surprise that the limits of detection which they derived from the uncertainty function defined by the reproducibility data were twenty-times higher than those determined under the within laboratory circumstances as a part of a method validation process.

4.1. A case study outline

The uncertainty functions known by now either tend to be entirely general and analyte, sample and analytical method independent, or in contrast to this focus on a particular analyte determined in a particular matrix with a particular method or predominantly so in the cases of proficiency trials.

In contrast to the cases which defined the Horwitz function and in which different analytes of different concentrations were determined in different matrices this research is sample orientated and tends towards the simultaneous ion chromatographic determination of the total ionic composition of a sample as completely as possible and with acceptable uncertainties. In the uncertainty budgets of analytical results obtained by the bottom-up approach the uncertainty associated with the interpolation of concentration from a calibration function equation can represent the major or one of the major contributions and this holds also for ion chromatography [7, 40].

The results of ion chromatographic analyses of seawaters and related samples can because of their wide concentration ranges between the minor and major ions provide an appropriate model for the consideration and evaluation if any general ion independent Horwitz-like prediction tool can be developed for the uncertainty of interpolation of concentrations from calibration lines under the conditions of determining simultaneously very different concentrations, and this was the objective of this research.

4.2. Materials and methods

The data set comprised 156 data points originating from the results of analyses of 25 natural and artificial sea waters, estuary waters and control standards. Fluoride, chloride, bromide, nitrate, sulfate, sodium, potassium, magnesium, calcium and strontium were determined simultaneously within two separate runs, one for anions and another for cations. Two different models of ion chromatographs were used. The factor of dilution for determination of anions was 16.7 for the estuary waters and 41.7 or 44.4 or 50.0 for all other samples. The factor of dilution for determination of cations was 50.0 for estuary water and 125 or 133 or 154 or 182 for other samples.

Anions were determined with a DIONEX DX-300 Gradient Chromatography System comprising: the Conductivity detector II, a 25 μL injection loop, an IonPac AS9-HC separation column with an AG9-HC pre-column and an ASRS-ultra II 4 mm suppressor used in a chemical mode of operation. The 9 mmol/L sodium carbonate eluent was used at the flow rate 1 mL/minute. The flow rate of the 25 mmol/L sulphuric acid regenerant solution was 7 mL/minute. ChromJet Integrator Spectra Physics Analytical was used for data acquisition.

Cations were determined with a DIONEX 500 chromatography system consisting of the IP 20 isocratic pump, the CD 20 conductivity detector, a 25 μL injection loop, an IonPac CS12A separation column with a CG12A pre-column and a CSRS-ultra II 4 mm suppressor used in a closed loop mode of operation at the 100 mA current setting. The eluent was 20 mmol/L methane sulphonic acid with flow rate 1 mL/minute. Data acquisition was performed with the PeakNet software.

A combined stock calibration standard with fluoride, chloride, nitrite, bromide, nitrate, phosphate and sulfate concentrations 2.00 mg/L, 19.0 g/L, 15.6 mg/L, 140 mg/L, 15.2 mg/L, 83.0 mg/L and 2.65 g/L respectively was prepared from the corresponding sodium salts, which were all p.a. grade chemicals dried for two hours at 110 C. Commercial 1 g/L phosphate ion standard solution purchased from Merck-Darmstadt (Germany) was used as a source of this ion. Five calibration solutions were prepared daily from the combined stock calibration standard. Calibration ranges for simultaneous determination of fluoride, chloride, nitrite, bromide, nitrate, phosphate and sulfate were: 0.01-0.08, 95-759, 0.078-0.624, 0.7-5.6, 0.076-0.608, 0.415-3.32, and 13.2-106 mg/L, respectively. Each calibration standard was injected once.

A combined stock calibration standard solution for the determination of cations was prepared from p.a. grade sodium chloride and strontium chloride dried for two hours at 110 C; sources of all other cations were commercial 1 g/L single cation standard solutions purchased from Merck-Darmstadt (Germany). The combined stock calibration standard solution contained sodium, potassium, calcium and strontium in concentrations 10.55 g/L, 382.1 mg/L, 400.3 mg/L, 13.00 mg/L respectively. Five calibration solutions were prepared daily from the combined stock calibration standard. The commercial 1 g/L magnesium ion standard solution was introduced directly into calibration solutions at this stage. Calibration ranges for simultaneous

http://dx.doi.org/10.5772/59012

determination of sodium, potassium, magnesium, calcium and strontium were: 26.4-132, 0.96-4.78, 3.15-15.8, 1.00-5.00 and 0.033-0.163 mg/L.

All solutions were prepared with distilled water additionally purified through the Milli-Q Academic system (Millipore, Billerica, MA, USA).

4.3. Results and discussion

Concentrations of ions (x_0) were obtained by interpolations from calibration functions using peak area as a measured variable (y_0) and by employing the ordinary least squares (OLS) linear regression model (equation 18). An intercept and a slope are denoted by a and b, respectively. The uncertainties (s_{x0}) associated with x_0 were calculated from equation 19. The symbols $s_{y/x}$, n, $\bar{y}$ and $\bar{x}$ stand for: a standard error of estimate, a number of calibration points, a mean of peak areas and a mean of concentrations of calibration solutions, respectively.

$$x_0 = (y_0 - a) / b \tag{18}$$

$$s_{x0} = \frac{s_{y/x}}{b}\sqrt{1 + \frac{1}{n} + \frac{(y_0 - \bar{y})^2}{b^2 \sum_i (x_i - \bar{x})^2}} \tag{19}$$

An appropriate way of expressing concentrations of ions relatively was necessary in order to test if any general ion independent Horwitz-like prediction tool can be developed. Expressing analytical results as mass fractions as it is usual in uncertainty functions is not appropriate for determinations of a total ionic composition.

A measure of the completeness of analysis in determining a total ionic composition of a sample is a cation-anion balance, requiring that the sums of equivalent concentrations of cations and anions are very similar (equation 20). An equivalent concentration c_{e_i}, (equation 21) takes into account the charge number of an ion z_i; nevertheless it is calculated from a mass concentration γ_i and a molar mass of an ion M_i similarly as amount concentration.

$$c_{e\_cation\_total} = \sum_{i=1}^{n} c_{e\_cation\_i} \quad c_{e\_anion\_total} = \sum_{i=1}^{n} c_{e\_anion_i} \tag{20}$$

$$c_{e\_i} = \frac{\gamma_i \times z_i}{M_i} \tag{21}$$

An established way of expressing ionic composition relatively is by using milliequivalent percent, meq% as a unit. Samples with similar meq% composition are analysed under similar experimental conditions. Results expressed in meq% for example help recognise a water type, a potential use, a hydrogeological origin, and processes in a water body. A fraction of an ion in a total cationic or anionic composition denoted by fraction_meq and expressed in meq% is obtained by applying equation 22.

$$\text{fraction\_meq} = \frac{c_{e\_cation}}{c_{e\_cation\_total}} \times 100 = \frac{c_{e\_anion}}{c_{e\_anion\_total}} \times 100 \tag{22}$$

Another widely used parameter related to determinations of ionic compositions of samples is TDS, total dissolved solids, calculated as a sum of mass concentrations (γ_i) of anions, cations and weak electrolytes; usually expressed in mg/L (equation 23). Mass concentrations of weak electrolytes usually contribute negligibly to TDS.

$$\text{TDS} = \sum_{i=1}^{n} \gamma_i \tag{23}$$

We used this parameter as a ground for an alternative way of expressing ionic composition relatively and introduced fraction_TDS expressed in percent in order to have a parameter expressed similarly as fraction_meq (equation 24).

$$\text{fraction\_TDS} = \gamma_{\text{ion}} / \text{TDS} \times 100 \tag{24}$$

Relative uncertainties of interpolations of concentrations of different ions from calibration lines related to their fractions in samples expressed in meq% are presented in figure 10. An unusual pattern was observed for calcium already when calibrations have been performed. The repeatability of the areas of chromatographic peaks obtained for calibration solution with the intermediate concentration was the worst and the least comparable between days, contributing to unequal uncertainties of interpolation assigned to similar concentrations of calcium in samples determined on different days. No obvious technical reason for these observations was recognised. A possible explanation lies in chemical equilibrium at the presence of sulfate in exceeding concentrations. Due to these observations calcium was excluded from further data evaluations.

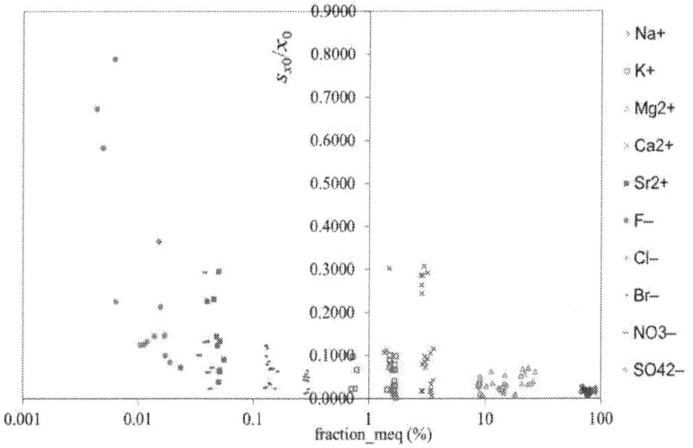

Figure 10. Relative standard uncertainties with which the mass concentrations of individual ions were determined from calibration lines, related to a fraction of an ion in the anionic or cationic composition of a sample, expressed in meq%

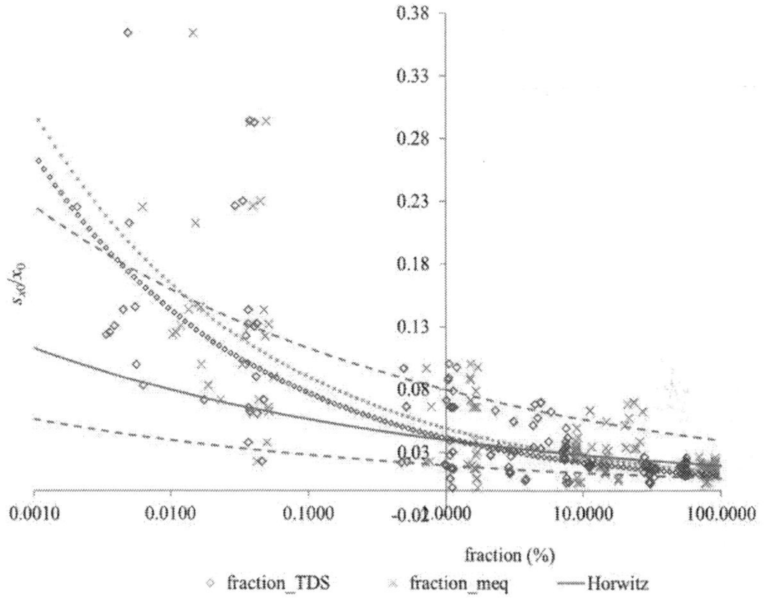

Figure 11. The Horwitz-like functions derived from the experimental results, and expressed either as a fraction of TDS (diamonds) or as a fraction of meq of anions or cations, meq% (crosses), compared with the Horwitz function (solid line) and its two-times and 0.5-times range (dashed lines)

The experimentally obtained data were linearized by applying logarithms to s_{x0}/x_0, fraction_TDS and fraction_meq. The OLS regression line equations 25 and 26 with the correlation coefficients, r 0.7875 and 0.7740 were derived, respectively. The rearrangements of these equations resulted in the Horwitz-like equations 27 and 28 and the corresponding curves are in figure 11 indicated with small size diamonds or crosses, respectively.

$$\log_{10}\left(\frac{s_{x0}}{x_0}\right) = -0.27002 \times \log_{10}(\text{fraction\_TDS}) - 1.38204 \tag{25}$$

$$\log_{10}\left(\frac{s_{x0}}{x_0}\right) = -0.26200 \times \log_{10}(\text{fraction\_meq}) - 1.30732 \tag{26}$$

$$\frac{s_{x0}}{x_0} = 0.041492 \times (\text{fraction\_TDS})^{-0.27002} \tag{27}$$

$$\frac{s_{x0}}{x_0} = 0.049281 \times (\text{fraction\_meq})^{-0.26200} \tag{28}$$

In order to enable the comparison of the experimentally obtained data with the equation of the Horwitz curve [15], which relates a percent relative standard deviation, $s_r(\%)$ and mass fraction, w (equation 15); we transformed this equation into equation 29, which relates the relative standard deviation, s_r and $w_{(\%)}$, which denotes a mass fraction of an analyte expressed in percent. The resulting curve is indicated in figure 11 with a solid line. Dashed lines indicate two-times and 0.5-times range of the Horwitz function.

$$s_r = 0.04\, w_{(\%)}^{-0.1505} \tag{29}$$

4.4. Lessons learned

In contrast to the observations derived from the data of interlaboratory comparisons that the Horwitz function holds well for mass fractions between 1.2×10^{-5} % and 13.8 % (equation 16) [34], the Horwitz-like curves are in the best agreement with the Horwitz function for fractions of ions between 1 % and 100 %, but deflect towards higher s_0/x_0 below 1 % as figure 11 demonstrates. This comes with no surprise since if a single minor component is targeted as it is usually a case with the interlaboratory comparisons a matrix elimination and pre-concentration are possible, in contrast to the ion chromatographic determination of the total ionic composition requiring that the minor and major ions are determined simultaneously what is a more demanding case.

5. An uncertainty prediction tool for determining simultaneously minor and major ions with ion chromatography

The results of ion chromatographic analyses of seawaters comprising 156 data serve as an appropriate model because of wide concentration ranges of ions. For two different ways of expressing relative fractions of ions in a sample the meaningful ion independent relations in a form of the Horwitz-like uncertainty functions can be confirmed. The prediction tools established through this research enable recognition of ions for which it can be expected that they are determinable simultaneously with ion chromatography with acceptable uncertainties. By setting appropriate criteria the prediction tools can be accustomed to the user defined requirements.

The meaningful ion independent curves in a form of Horwitz-like functions are presented in figure 12 for the entire data range together with their 0.5-times to two-times, and 0.33-times to three-times ranges which are indicated with dashed lines. In both figures nearly all data points lie in the 0.33-times to three-times range, with the fraction_TDS representation having fewer points outside this range.

These figures can in association with equations 27 and 28 serve as prediction tools. If for example we set a criterion that for any single determination of a concentration the relative uncertainty of interpolation of a concentration from a calibration line equation greater than 0.33 is considered unacceptable ions present in a sample in a fraction_TDS below 0.027 % or in a fraction_meq below 0.046 % cannot be successfully determined simultaneously with the others, as vertical solid lines in figure 12 indicate. In prospective considerations different

http://dx.doi.org/10.5772/59012

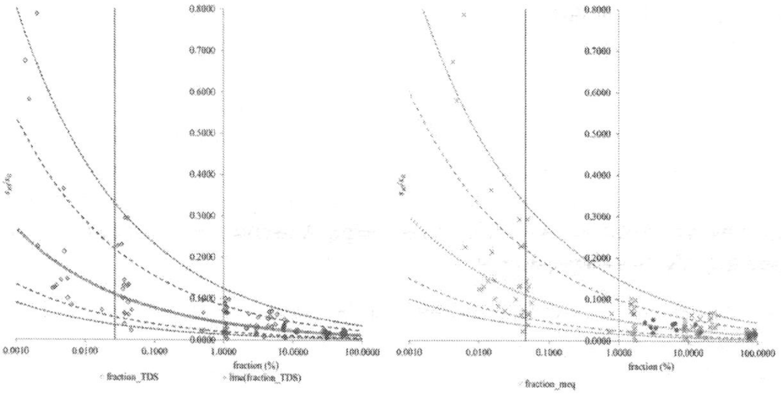

Figure 12. A Horwitz-like function for the results expressed as a TDS fraction, left or in meq%. right, and a three-times, two-times, 0.5-times and 0.33-times ranges (dashed lines)

criteria can be set for the relative uncertainty of interpolation of a concentration from a calibration line equation and decisions made if it is reasonable trying to determine a particular ion in a sample simultaneously with the others.

An independent confirmation of the prediction tools presented here is not an easy task, firstly due to a lack of studies comprising determinations of the total ionic composition of samples in such a wide concentration range, and secondly if this first condition is fulfilled at least for a narrower concentration range extending over two to three orders of magnitude, the uncertainty values are seldom given or evaluated in a way that would allow for a comparison.

To the best of our knowledge the research of Mosello *et al* comes the closest to fulfilling both criteria [41]. For the AQUACON interlaboratory exercise, which evaluated the ion chromatographic performance, the expected values and uncertainties were determined by two reference laboratories and given for determination of the main anions and cations in two freshwater samples. The data, which are presented in figure 12, right, with dots, are in a good agreement with the Horwitz-like function. All data points are within a 0.5-times to two-times range.

6. Conclusions

Determination of total ionic composition of water samples is an important aspect of water quality evaluation. The ordinary least squares linear regression method is most frequently used in instrumental analysis. In spite of the fact that uncertainty associated with the concentration of ions in calibration solutions is considered negligible if compared with the uncertainty associated with a measured quantity, the uncertainty associated with an analytical result obtained with interpolation from a calibration line equation and taking into account the uncertainty originating from a dilution of a sample has to be combined at the end also with the uncertainty associated with the concentration of the ion in a stock calibration standard solution.

This research confirmed that in preparation of multi-ion standard solutions that are intended for analyses of water sample in which concentrations of ions extend over wider concentration ranges the impurities originating from other chemical have to be taken into account in evaluations of concentrations of ions and standard uncertainties associated with them. Otherwise standard uncertainties can be underestimated. This is of special importance when it comes to high purity of a prime chemical from which the ion under consideration originates, higher mass fraction of this ion present as an impurity in another chemical and higher mass ratios of the matrix ion and ion under consideration. A refined and simplified methodology for further evaluations is suggested.

The Horwitz-like functions established through this research enable recognition of ions for which it can be expected that they are determinable simultaneously with ion chromatography with an acceptable uncertainty if their approximate relative fractions in a sample were previously estimated. The prediction tools suggested here in spite of the fact of being derived from the experimental data using a particular calibration scheme and data treatment can serve as a useful guide which can be adapted to user defined requirements by setting appropriate criteria.

Author details

Nataša Gros*

Address all correspondence to: natasa.gros@fkkt.uni-lj.si

Faculty of Chemistry and Chemical Technology, University of Ljubljana, Ljubljana, Slovenia

References

[1] Chapman D. Water quality assessments a guide to the use of biota, sediments and water in environmental monitoring New York: E & F Spon; 1996. Available from: http://www.who.int/water_sanitation_health/resourcesquality/wqa/en/.

[2] Hounslow AW. Water Quality Data: Analysis and Interpretation. Boca Raton, New York: Lewis Publishers; 1995.

[3] Gros N, Camőes MF, Oliveira C, Silva MCR. Ionic composition of seawaters and derived saline solutions determined by ion chromatography and its relation to other water quality parameters. Journal of Chromatography A. 2008;1210(1):92-8.

[4] Prichard E, Barwick V. Quality Assurance in Analytical Chemistry. Chichester: Wiley; 2007.

[5] Gonzalez AG, Herrador MA. A practical guide to analytical method validation, including measurement uncertainty and accuracy profiles. Trac-Trends in Analytical Chemistry. 2007 Mar;26(3):227-38. PubMed PMID: ISI:000245499300019. English.

[6] International Organization for Standardization, Guide to the expression of uncertainty in measurement. Geneve, Switzerland1995.

[7] Ellison SLR, Williams A. Eurachem/CITAC guide: Quantifying Uncertainty in Analytical Measurement. Third ed2012.

[8] Meyer VR. Measurement uncertainty. Journal of Chromatography A. 2007 Jul; 1158(1-2):15-24. PubMed PMID: ISI:000248418500003. English.

[9] Kragten J. Calculating Standard Deviations and Confidence-Intervals with a Universally Applicable Spreadsheet Technique. Analyst. 1994 Oct;119(10):2161-5. PubMed PMID: WOS:A1994PN35600003.

[10] Chew G, Walczyk T. A Monte Carlo approach for estimating measurement uncertainty using standard spreadsheet software. Anal Bioanal Chem. 2012 Mar;402(7): 2463-9. PubMed PMID: WOS:000300454900026.

[11] Gros N. The comparison between Slovene and Central European mineral and thermal waters. Acta Chim Slov. 2003;50(1):57-66. PubMed PMID: WOS: 000181815100005.

[12] Michalski R. Ion chromatography as a reference method for determination of inorganic ions in water and wastewater. Critical Reviews in Analytical Chemistry. 2006 Apr-Jun;36(2):107-27. PubMed PMID: ISI:000238283300004.

[13] Gros N, Gorenc B. Performance of ion chromatography in the determination of anions and cations in various natural waters with elevated mineralization. Journal of Chromatography A. 1997 May;770(1-2):119-24. PubMed PMID: ISI:A1997XC91100016.

[14] Gros N. Ion Chromatographic Analyses of Sea Waters, Brines and Related Samples. Water. 2013 Jun;5(2):659-76. PubMed PMID: WOS:000320769800020.

[15] Horwitz W. Evaluation of analytical methods used for regulation of foods and drugs. Anal Chem. 1982;54(1):A67-&. PubMed PMID: WOS:A1982MW08100004.

[16] Horwitz W, Albert R. The Horwitz ratio (HorRat): A useful index of method performance with respect to precision. Journal of Aoac International. 2006 Jul-Aug;89(4): 1095-109. PubMed PMID: WOS:000239470600028.

[17] Kolb M, Hippich S. Uncertainty in chemical analysis for the example of determination of caffeine in coffee. Accredit Qual Assur. 2005 May;10(5):214-8. PubMed PMID: ISI:000229510400005. English.

[18] Gros N, Camoes MF, da Silva R. Detailed uncertainty budget for major and minor ions in stock combined calibration standards-Influence of impurities in chemicals.

Analytica Chimica Acta. 2010 Feb;659(1-2):85-92. PubMed PMID: WOS: 000274046900009.

[19] Carlé W. DIE Mineral-und Thermalwässer von Mitteleuropa, Geologie, Chemismus, Genese. Stuttgart: Wissenschaftliche Verlagsgesellschaft MBH; 1975.

[20] van Look G, Meyer VR. The purity of laboratory chemicals with regard to measurement uncertainty. Analyst. 2002 Jun;127(6):825-9. PubMed PMID: ISI: 000176396800023.

[21] Jauhiainen T, Moore J, Peramaki P, Derome J, Derome K. Simple procedure for ion chromatographic determination of anions and cations at trace levels in ice core samples. Analytica Chimica Acta. 1999 May;389(1-3):21-9. PubMed PMID: ISI: 000080527900003. English.

[22] Saitoh K, Sera K, Shirai T, Sato T, Odaka M. Determination of elemental and ionic compositions for diesel exhaust particles by particle induced X-ray emission and ion chromatography analysis. Analytical Sciences. 2003 Apr;19(4):525-8. PubMed PMID: WOS:000182434800007.

[23] Nakatani N, Kozaki D, Mori M, Tanaka K. Recent Progress and Applications of Ion-exclusion/Ion-exchange Chromatography for Simultaneous Determination of Inorganic Anions and Cations. Analytical Sciences. 2012 Sep;28(9):845-52. PubMed PMID: WOS:000309333000003.

[24] Gros N, Gorenc B. Simple recognition of similar samples for the ion-chromatographic determination of the main cations. Journal of Chromatography A. 1997 Nov;789(1-2): 323-7. PubMed PMID: WOS:000070924700029.

[25] Horwitz W, Kamps LR, Boyer KW. Quality Assurance in the Analysis of Foods for Trace Constituents. Journal of the Association of Official Analytical Chemists. 1980;63(6):1344-54. PubMed PMID: WOS:A1980KR83000036.

[26] Barwick VJ, Ellison SLR. Estimating measurement uncertainty using a cause and effect and reconciliation approach Part 2. Measurement uncertainty estimates compared with collaborative trial expectation. Analytical Communications. 1998 Nov; 35(11):377-83. PubMed PMID: WOS:000077005900007.

[27] Chen L, Eitenmiller RR. Single laboratory method performance evaluation for the analysis of total food folate by trienzyme extraction and microplate assay. Journal of Food Science. 2007 Jun-Jul;72(5):C243-C7. PubMed PMID: WOS:000247780200009.

[28] Noel L, Carl M, Vastel C, Guerin T. Determination of sodium, potassium, calcium and magnesium content in milk products by flame atomic absorption spectrometry (FAAS): A joint ISO/IDF collaborative study. International Dairy Journal. 2008 Sep; 18(9):899-904. PubMed PMID: WOS:000258432800003.

[29] Boyer KW, Horwitz W, Albert R. Interlaboratory Variability in Trace-Element Analysis. Anal Chem. 1985;57(2):454-9. PubMed PMID: WOS:A1985AAV7000018.

[30] Albert R, Horwitz W. A heuristic derivation of the Horwitz curve. Anal Chem. 1997 Feb;69(4):789-90. PubMed PMID: WOS:A1997WH35700039.

[31] Thompson M, Lowthian PJ. A Horwitz-like Function Describes Precision in a Proficiency Test. Analyst. 1995 Feb;120(2):271-2. PubMed PMID: WOS:A1995QH46300008.

[32] Arnaud J, Jones RL, LeBlanc A, Lee MY, Mazarrasa O, Parsons P, et al. Criteria to define the standard deviation for proficiency assessment for the determination of essential trace elements in serum: comparison of Z-scores based on the Horwitz function or on biological variability. Accredit Qual Assur. 2009 Aug;14(8-9):427-30. PubMed PMID: WOS:000268949400004.

[33] Thompson M, Lowthian PJ. The Horwitz function revisited. Journal of Aoac International. 1997 May-Jun;80(3):676-9. PubMed PMID: WOS:A1997XA02300027.

[34] Thompson M. Recent trends in inter-laboratory precision at ppb and sub-ppb concentrations in relation to fitness for purpose criteria in proficiency testing. Analyst. 2000;125(3):385-6. PubMed PMID: WOS:000085594500001.

[35] Linsinger TPJ, Josephs RD. Limitations of the application of the Horwitz equation. Trac-Trends in Analytical Chemistry. 2006 Dec;25(11):1125-30. PubMed PMID: WOS: 000243249800019.

[36] Thompson M. Limitations of the application of the Horwitz Equation: A rebuttal. Trac-Trends in Analytical Chemistry. 2007 Jul-Aug;26(7):659-61. PubMed PMID: WOS:000249516400012.

[37] Thompson M. Uncertainty functions, a compact way of summarising or specifying the behaviour of analytical systems. Trac-Trends in Analytical Chemistry. 2011 Jul-Aug;30(7):1168-75. PubMed PMID: WOS:000294318200029.

[38] Thompson M. The Characteristic Function, a Method-Specific Alternative to the Horwitz Function. Journal of Aoac International. 2012 Nov-Dec;95(6):1803-6. PubMed PMID: WOS:000313142800035.

[39] Cote I, Robouch P, Robouch B, Bisson D, Gamache P, LeBlanc A, et al. Determination of the standard deviation for proficiency assessment from past participant's performances. Accredit Qual Assur. 2012 Aug;17(4):389-93. PubMed PMID: WOS: 000306948200008.

[40] Gros N, Camoes MF, da Silva R. Uncertainty budget for simultaneous determination of minor and major ions in seawater with ion chromatography confronted with uncertainties in concentrations of calibration standards. Anal Lett. 2010;43(7-8):1317-29. PubMed PMID: WOS:000277838400024.

[41] Mosello R, Tartari GA, Marchetto A, Polesello S, Bianchi M, Muntau H. Ion chromatography performances evaluated from the third AQUACON freshwater analysis interlaboratory exercise. Accredit Qual Assur. 2004 Mar;9(4-5):242-6. PubMed PMID: ISI:000189285700012. English.

Chapter 6

Statistical Modelling of Water Quality Time Series – The River Vouga Basin Case Study

Marco André da Silva Costa and
Magda Sofia Valério Monteiro

Additional information is available at the end of the chapter

http://dx.doi.org/10.5772/59814

Introduction

"Water is the principle of all things" Thales of Miletus.

Water quality models are essential tools for the assessment of the impact of ecosystem changes in response to variable inputs, as well as the interactions occurring within the system [1]. Taking this into account, developed countries have continually been monitoring and classifying their water. The over-abstraction of fresh water is a problem in certain parts of Europe, especially in coastal countries and islands in the Mediterranean, leading to the decrease of groundwater bodies, the deterioration of water quality and the destruction of certain habitats [2]. Water quality monitoring procedures may be used in decision-making processes for supporting policy options. For this reason, several European Union (EU) countries have developed a national water quality system considering the characteristic structures of their own rivers and have used these types of indicators to evaluate their current water quality levels. Water quality monitoring is usually conducted using physical and chemical parameters; nevertheless, biological factors such as algae, especially diatoms, benthic macro invertebrates and fish, have recently been used for this purpose [3]. Nowadays, the majority of the population in northern and central Europe is connected to wastewater treatment plants that make use of advanced treatments. This development allows for a decrease in the load of organic matter and nutrients transported by rivers and subsequently lead to an improvement of the state of water resources [4].

The management of water resources is regulated by EU directives and their transposition into national legislation. For instance, in Portugal, the Law nº58/2005 (Law of Water) ensures implementation into national law the Directive nº2000/60/CE (the Water Framework Directive,

WFD), which creates the institutional framework for the sustainable management of surface, interior, transitional, coastal and even groundwater. According to this directive, each member-state has to project, improve and recover all surface waters in order to achieve the good qualitative and quantitative status of all water bodies by 2015 [5,6]. The Decree-Law nº 77/2006 complements the WFD by characterizing the waters of a river basin. A regulatory instrument establishes the status of surface waters and groundwater and their ecological potential.

According to the directive's framework, the monitoring of water resources essentially has two purposes. The first is the evaluation of water status (surveillance monitoring) and the second is the implementation of programmes that include measures for identifying water resources at risk of failing environmental objectives (operational monitoring). If surveillance-monitoring reveals that the water in question has not reached the necessary "ecological status", a basin management plan should be implemented for the duration of one year. During this process, indicative parameters of all elements of biological, hydromorphological and general physico-chemical quality, as well as significant discharges in the basin containing pollutants, should be monitored in order to permit the classification of ecological status. In operational monitoring, measures are implemented and the status of water is evaluated as a result of the implementation of such measures [7].

In this context, every European country organizes itself using various entities that monitor water resources at various levels. The water area in Portugal consists of six large sets of entities: the trusteeship policy (national and regional), advisory bodies (national and regional), public water management (national and regional), utilities users and regulators (urban, agricultural, energy developments and multipurpose), mixed structures and associations and non-governmental entities.

The Water Portal website aims to respond to the provisions of Chapter VIII of the Water Law, especially its art. 87º, which empowers the National Water Authority – Institute of Water (Instituto da Água I.P.) to create a national information system for water and place it at the disposal "of... entities that have responsibilities [in] exercising public functions or [in] providing public services directly or indirectly related to water [to] the technical and scientific community and the public in general." [8]. The national water resources monitoring system is supported by a database that stores, prepares and publicize both hydro-meteorological and water quality (surface and groundwater) data, which have been collected by the monitoring network of the Ministry of the Environment (through the http://snirh.pt portal system). The National Information System for Water Resources (Sistema Nacional de Informação de Recursos Hídricos – SNIRH) was created by the Institute of Water in mid-1995. The monitoring network consists of automatic and conventional stations, some of which are equipped with remote transmission. The portal system also publishes monthly thematic syntheses aimed at characterizing the national water availability, technical reports, the mapping of water resources (e.g., flood zones) and maintaining technical documents and photographs related to water resources [9], following the significant economic support from European funds devoted to investments in sewage systems. Portugal has had slow growth in the area of wastewater treatment. The rate of the population connected to public wastewater treatment plants was about 56% in 2005 and 74% in 2009 (see Figure 1).

http://dx.doi.org/10.5772/59814

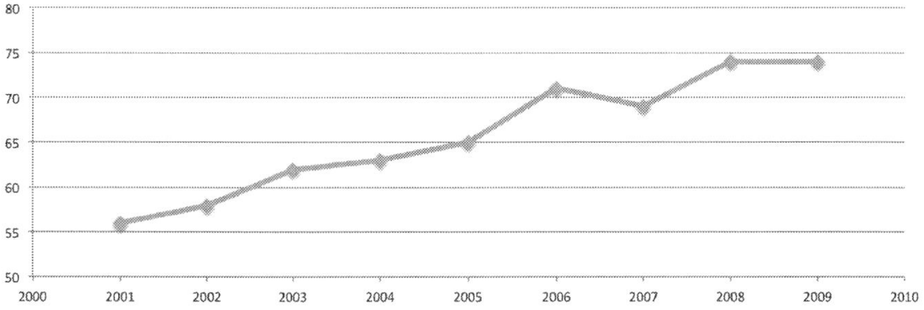

Figure 1. Evolution of the percentage of the population connected to public wastewater treatment plants in mainland Portugal between 2001 and 2009 [10].

Despite high investments at various levels of Portuguese public administration and those of economic actors, the economic crisis of recent years has led to a disinvestment in the monitoring and in the water treatment systems. Undoubtedly, the economic constraints of local authorities in Portugal have led to a reduction in investments in waste management, which may jeopardize the quality of surface water in several Portuguese hydrological basins (see Figure 2).

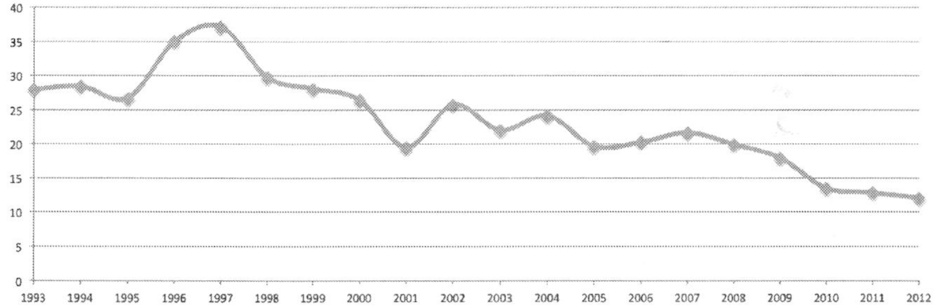

Figure 2. Investments in municipal waste management, in millions of euros, between 1993 and 2012 in Portugal [10].

Surface water-monitoring systems are important tools for analysing environmental processes and for identifying changes within these processes. In this context, statistical tools and methodologies are relevant for assessing water quality and for allowing the analysis of available data yielded by information systems. The temporal evolution analysis of water quality indicators is an important tool in this context, as the real-time monitoring procedures. Thus, modern statistical techniques are an integral part of any water monitoring process.

This paper presents a set of issues that analysts encounter in the analysis of data that support decision-making processes. The focus of this work is on contributions to the analysis of time series of water quality variables. The study of these time series has characteristics related to data collection procedures, as well as to several economic and geographic aspects. Some characteristics of water quality variables pose challenges to their statistical modelling, namely, seasonality, change-point detection, the existence of outliers, etc. Some of these topics will be addressed in this chapter. The presentation of these topics is accompanied by the study of water quality data from the hydrological basin of the Vouga River and the Ria de Aveiro lagoon. In particular, statistical modelling is performed using data pertaining to the dissolved oxygen concentration variable (in *mg/l*) from the Carvoeiro water-monitoring site in the Vouga River basin. Section 2 discusses the area under study, while Section 3 presents a data description with an exploratory analysis of the primary monitoring sites of the Vouga basin. Section 4 addresses the analysis of a time series through the decomposition into its main components and their treatment, while Section 5 presents the modelling of time series through a state space model approach.

2. The study area

The Vouga River is situated at the centre of Portugal at an altitude of about 930m, near the geodesic landmark Facho da Lapa in Serra da Lapa, a mountain located in the district of Viseu. The river flows 148 km before empting into Ria de Aveiro. The watershed of the Vouga (also referred to as Vouga e Ribeiras Costeiras) is the second largest basin among the watercourses that run exclusively in Portuguese territory, comprising a total area of 3706 Km^2. More specifically, the Vouga basin is located in the transition zone between the north and south of Portugal, i.e., between the watersheds of the Douro in the north and the Mondego in the south (see Figure 3).

Several rivers flow into the Ria de Aveiro estuarine system. The hydrologic regime involves a summer low flow condition and the dynamics of the coastal lagoon are dominated by tidal oscillation. Ria de Aveiro is characterized by its rich biodiversity, as well as by increasing pressure related to anthropogenic activities near its margins, i.e., building and land occupation and agricultural and industrial activities. This has resulted in a significant change to the lagoon's morphology and in the constant input of a large volume of anthropogenic nutrients, as well as contaminant loads, which in turn has had a negative impact on water circulation and the water quality of the lagoon, [11]. The construction, management and operation of the multi-municipality system drainage of the Ria de Aveiro is the responsibility of SIMRIA – Integrated Sanitation of Municipalities of Ria, SA, a private company with a public capital majority (established by Decree-Law nº 101/97 of 26 April).

Several studies have been developed around this watershed and in particular, around the Ria de Aveiro lagoon. Some studies focus on the ecological systems related to Ria de Aveiro and the diversity of its flora and fauna (e.g., [12, 13]. Other studies have examined the biochemical properties of the region's environmental systems. Still others have studied the hydro-meteorological aspects of the watershed. This diversity and multidisciplinary approach to the Ria de Aveiro lagoon and its associated watershed was recently formalized by the University of Aveiro as '*Grupo uariadeaveiro*'. The group aims to monitor and contribute to the management

http://dx.doi.org/10.5772/59814

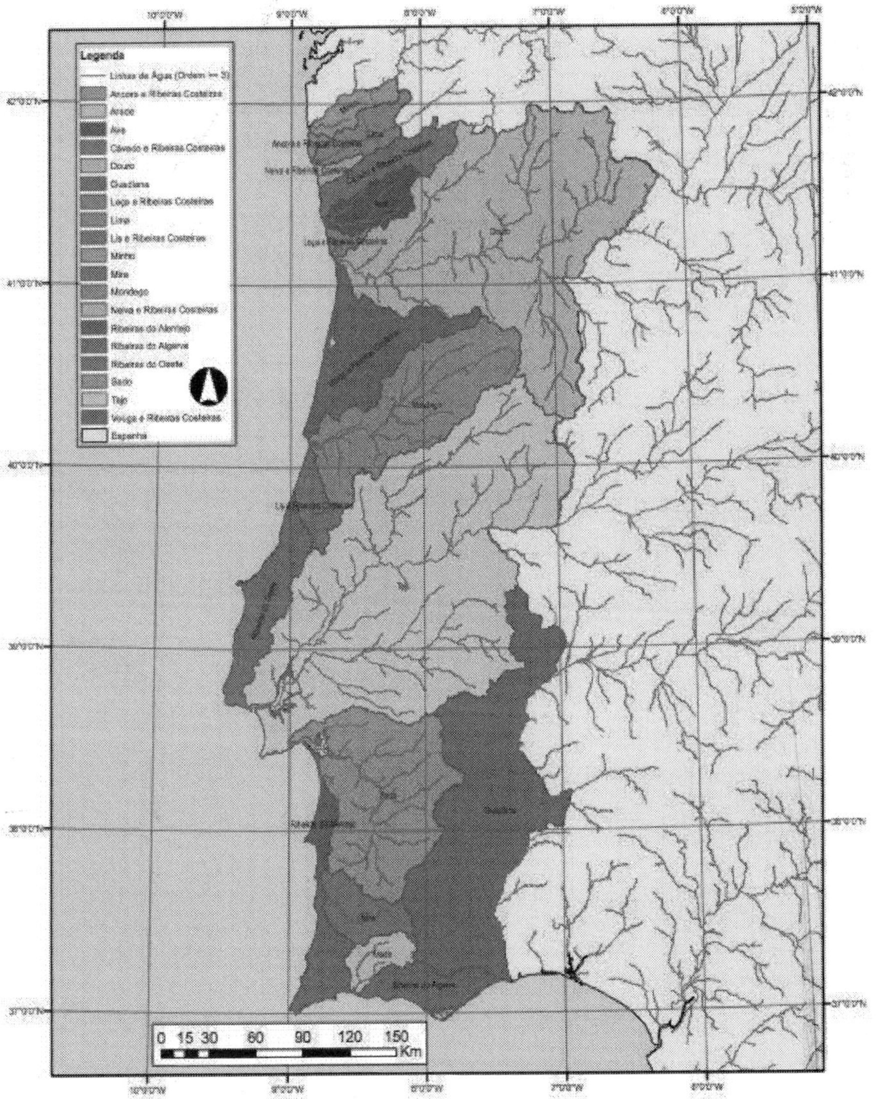

Figure 3. Hydrological basin of mainland Portugal [9].

of the Ria de Aveiro by focusing mainly on knowledge transfer activities; by doing so, it can also contribute to the enhancement of other activities within the University of Aveiro. Additionally, special attention to the activities of the primary entities and relevant stakeholders associated with the management of the Ria de Aveiro lagoon can in this way be maintained in order to ensure the useful and effective contribution of the University of Aveiro. The University of Aveiro, through its *uariadeaveiro* group, is committed to facilitating the provision of scientific knowledge about this lagoon area and places it at the service of the region. Thus, the University of Aveiro hopes to contribute to the improvement of the protection, enhancement and management of the Ria de Aveiro.

The website http://www.ua.pt/riadeaveiro/ (see Figure 4) was established as a mechanism for the dissemination of knowledge and activities of interest pertaining to the Ria de Aveiro. This work in progress aims to establish itself as the most important collection of data about the Ria de Aveiro in the country. It is hoped that this initiative will help to create internal synergies within the scientific community focused on the Ria de Aveiro and also to strengthen bridges of communication with regional stakeholders in order to maximize the successful management of the Ria.

Figure 4. The website http://www.ua.pt/riadeaveiro/ located within the University of Aveiro site (http://www.ua.pt).

The webpage of the *uariadeaveiro* group includes five menus. The first is dedicated to the presentation of the group and its goals. The second is dedicated to the library, where access can be gained to all scientific and non-scientific publications regarding the Ria de Aveiro. The third page provides spatial information on the Ria de Aveiro and allows for access to other pages with geographic information of interest regarding the Ria systems. The fourth page is devoted to the approximation entities and users of the Ria. The fifth page discusses research projects completed and ongoing, relevant training courses and activities related to the Ria de Aveiro [14].

The average fresh water flow into the Ria de Aveiro is about 40 m^3/s. The Vouga and Antuã rivers are the main sources of fresh water, with an average annual flow of 24 m^3/s and 2.4 m^3/s; both of these rivers are part of the Vouga watershed, [15].

The main tributaries of the Vouga River are, from upstream to downstream, the River Mel, the Sul River, the Varoso, the River Teixeira, the River Arões, the River Mau and the Caima River on the right bank; on its left bank, the River Ribamá, the Marnel and the River Águeda with its major tributary, the Alfusqueiro. Officially, the catchment of the River Vouga consists in the rivers that dewater into the Ria de Aveiro with the exception of Vala da Fervença, in the Mira area, which flows to the southern end of the Ria de Aveiro [16].

A set of water monitoring sites (see Figure 5), where several variables pertaining to water quality can be collected, is available in the hydrological basin of the Vouga River. However, some problems have arisen in the context of statistical modelling, i.e., some water monitoring sites and variables yield little data and/or have missing values. Furthermore, due to a lack of economic resources and other factors, data collection was discontinued at some sites.

Within the SNIRH system, 78 water-monitoring sites are registered within the hydrological basin of the Vouga River. However, data collection is not continuous and some stations had been deactivated somewhere in time. Relative to the DO concentration, 26 stations showed a significant data set, with data up to May 2013 (the last month available in the system). Hence, the statistical analysis will be focus mainly on this data set.

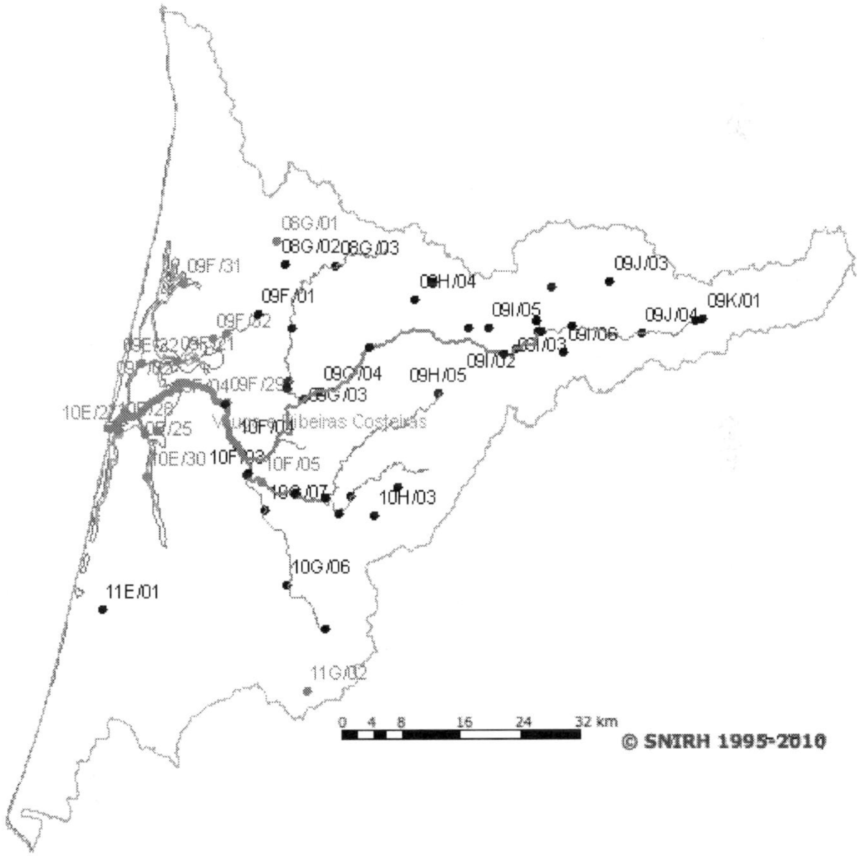

Figure 5. Water monitoring sites locations (red – inactive in the present; black – actives in the present; green – hydrological basin geographical boundaries), [9].

3. Data set description

The statistical analysis was performed using a data set related to the dissolved oxygen concentration in a large set of water monitoring sites located in the hydrological basin of the River Vouga, a basin associated with the Ria de Aveiro lagoon. The analysis focused on the DO concentration because it is one of the most important variables in the evaluation of a river's water quality [6] and because of its continuity in terms of measurements taken at all selected water quality monitoring sites under analysis.

The DO variable is widely recognized as one of the most relevant for the monitoring of water quality. DO concentration is controlled by several physical processes such as the rate of production of oxygen by algae, the nitrogen cycle and other biochemical processes [17]. At the same time, available oxygen is the primary factor affecting aquatic life in rivers. A good level of dissolved oxygen is essential for aquatic life. Oxygen concentration is therefore, the most important parameter for aquatic life and ecological states in estuaries [11]. Dissolved oxygen analysis measures the amount of gaseous oxygen dissolved in an aqueous solution. Oxygen gets into water by diffusion from the surrounding air, by aeration and as a waste product of photosynthesis. Adequate dissolved oxygen is necessary for good water quality and is a necessary element to all forms of life. Natural stream purification processes require adequate oxygen levels in order to provide for aerobic life forms. When dissolved oxygen levels in water drop below 5 mg/l, aquatic life is put under stress. With a decrease in oxygen concentration the stress on life forms increases. Oxygen levels that remain below 1-2 mg/l for a few hours can result in large numbers of fish deaths [18].

The SNIRH system holds data regarding the DO concentration of 78 water-monitoring sites between April 1989 and May 2013. However, among them several sites hold little or exhibit several missing values, which are associated with large gaps between measurements. Thus, the statistical descriptive analysis will be performed primarily on the largest time series. A significant number of locations had less than 23 observations for all time periods. These time series will not be considered in the descriptive analysis. On the one hand, only 26 water-monitoring sites had at least 100 observations and these will be considered in the exploratory analysis. On the other hand, only between January 2002 and May 2013 are consistent data sets available for all locations. Thus, the main data set that will be considered consist of these 26 time series during this period.

Data collection was not equally distributed during all time periods. Sometimes there was more than one measurement in a month. Therefore, in order to not lose information, we took the average of all measurements if there was more than one observation in a month. Table 1 presents descriptive statistics for the 26 time series between January 2002 and May 2013 (137 months) and Figure 6 shows the 26 time series representations. The graphical representation clearly shows that time series exhibited seasonal behaviour, as was expected due to the nature of the data.

http://dx.doi.org/10.5772/59814

Site	Code	Abbrev.	N. obs.	Min	Max	Average	St dev	Skew	Kurtosis
Agadão	10H/03	AGA	111	5.8	11.0	8.74	1.26	-.35	-.25
Carvoeiro	09G/03	CAR	112	6.2	11.0	8.79	1.18	-.15	-.40
Alombada	09G/04	ALO	113	6.1	11.0	8.90	1.08	-.51	.41
Captação Burgães	08G/03	BUR	122	6.5	12.6	9.40	1.16	-.15	-.18
Captação Rio Ínsua	08G/02	INS	122	6.4	12.4	9.31	1.05	-.33	.65
Ponte Redonda	10G/05	RED	112	4.6	11.5	8.88	1.22	-.47	.70
Frossos	09F/04	FRO	110	4.5	11.0	8.17	1.22	-.16	-.24
Pampilhosa	11G/02	PAM	100	4.3	12.0	7.95	1.68	.03	-.35
Ponte São João de Loure	10F/04	LOU	112	5.4	11.0	8.24	1.25	-.02	.01
Ponte Vale Maior	09G/01	MAI	112	6.2	12.0	8.62	1.12	-.05	-.05
Ponte Águeda	10G/02	AGU	111	5.1	11.0	8.39	1.20	-.34	-.18
São Tomé	11E/01	TOM	118	5.0	11.0	7.88	1.16	-.07	-.37
Aç. Maeira	09K/01	MAE	115	5.6	11.0	8.50	1.20	-.16	-.41
Aç. Rio Alfusqueiro	09H/05	ALF	113	2.9	12.0	7.80	1.75	-.31	-.01
Pindelo Milagres	09J/03	MIL	110	4.6	12.0	8.16	1.42	-.14	-.09
Ponte Antim	09I/05	ANT	113	0.8	12.0	7.38	2.05	-.68	.14
Ponte Pouves	09I/03	POU	115	2.6	11.0	8.27	1.46	-.67	1.30
Ponte Vouzela	09I/02	VOZ	109	1.8	13.0	8.10	1.91	-.84	1.32
São João Serra	09H/04	SER	115	6.0	12.0	8.70	1.18	.01	-.08
São Miguel Mato	09I/06	MAT	111	4.3	12.0	8.44	1.53	-.41	.03
Vouguinha	09J/04	VOG	114	5.4	11.0	8.42	1.35	-.26	-.52
Estarreja	09F/05	EST	114	3.4	11.0	7.62	1.32	-.71	1.02
Perrães	10G/07	PER	111	4.6	9.8	7.19	1.17	.14	-.53
Ponte Canha (Vouga)	10G/06	CAN	114	2.6	10.1	6.89	1.92	-.26	-.96
Ponte Minhoteira	09F/01	MIN	112	0.7	10.0	7.73	1.50	-1.10	3.60
Ponte Requeixo	10F/03	REQ	111	3.9	11.0	7.13	1.52	.33	-.09

Table 1. Descriptive statistics for dissolved oxygen concentration between January 2002 and May 2013.

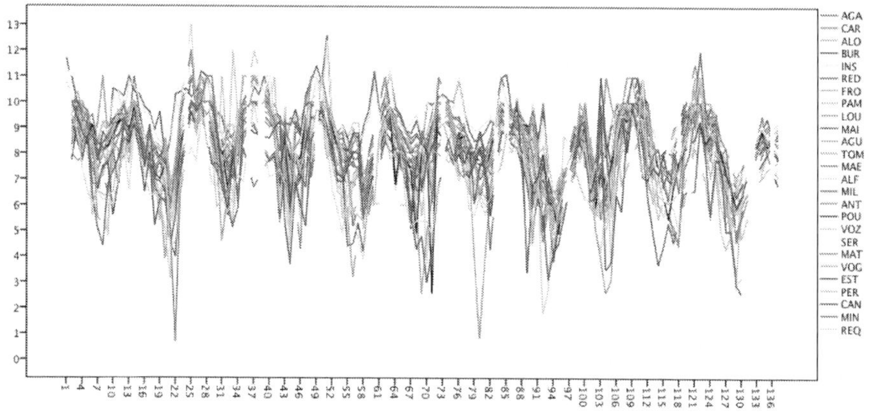

Figure 6. Time series for dissolved oxygen concentration in the 26 water-monitoring sites from January 2002 to May 2013.

An exploratory analysis showed that, in general, observations were not normally distributed. Indeed, in some locations, the distribution of observations is skewed or leptokurtic. This fact must be considered in the modelling procedures, since Gaussian distribution is a usual assumption in several statistical analyses. Figure 7 represents box-plots for DO concentrations. Boxplots are able to identify several moderate outliers in many locations. However, due to existence of temporal correlation structure in the measurements, the analyses of outliers must be carefully considered.

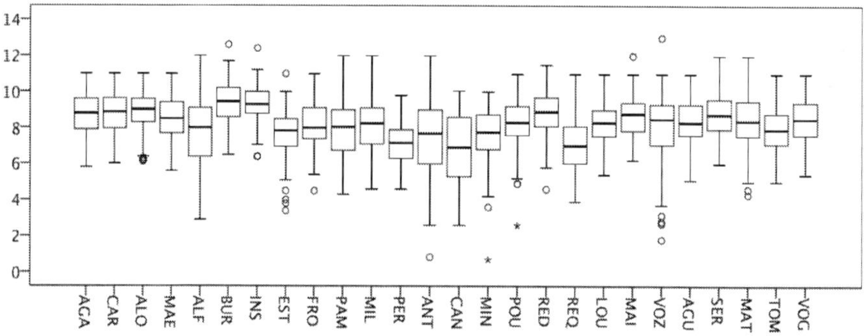

Figure 7. Boxplots for dissolved oxygen concentration from January 2002 to May 2013.

As DO concentration exhibits seasonal behaviour, it is necessary to analyse the monthly averages. Figure 8 shows the monthly averages of DO concentration during the analysed period for the 26 water-monitoring sites. These results indicated that DO concentration was greater in the winter months and presented lower values in the summer months. This fact is directly related to hydro-meteorological conditions, since DO concentration is influenced primarily by precipitation amounts and temperature. The monthly standard deviations presented in Figure 9 exhibit a diffuse pattern. Although it is difficult to identify a common pattern for all sites, a graphic representation indicates that in several locations there was larger variability during the months of January, September, October and November. The lower variability of the monthly DO concentration measures in December may be justified by the reduced observations typically collected during this month.

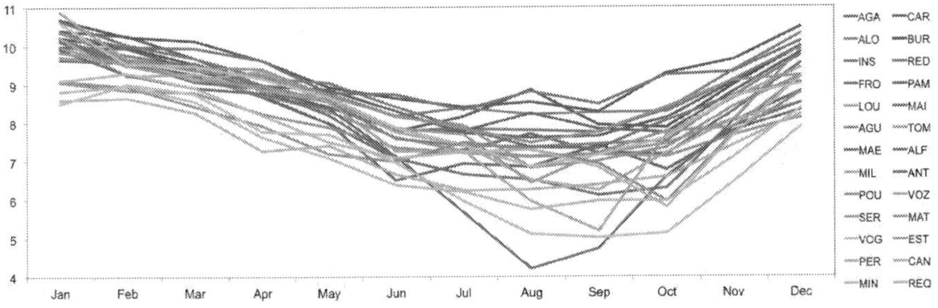

Figure 8. Monthly averages of DO concentration for the 26 water-monitoring sites between January 2002 and May 2013.

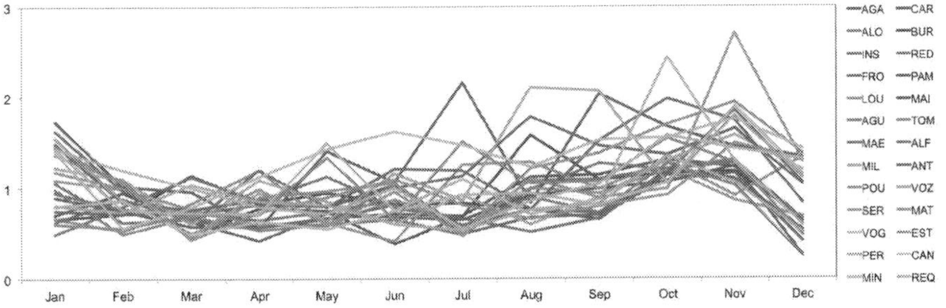

Figure 9. Monthly standard deviations in DO concentration for the 26 water-monitoring sites between January 2002 and May 2013.

4. Time series analysis

This section addresses several issues in the statistical modelling of time series water quality variables. Water quality variables present characteristics that pose some challenges to modelling procedures. By their nature, these variables have properties that must be incorporated in order for statistical techniques to yield good performance.

There are different ways of dealing with these features based on different approaches. One way of modelling a time series with components as either a trend or a seasonal/stochastic behaviour is by employing the time series decomposition method. This method consists of sequentially and individually modelling each component so that trends and/or seasonal components can be estimated and subtracted from the data to generate a noise sequence.

In the present work, we present, from a practical point of view, the most common methodologies applied to the time series of DO concentration in the Carvoeiro water-monitoring site. The choice of this monitoring site is related to its location in the main river, in the middle section, near to a point of abstraction of water for human consumption. Furthermore, the time series data in this location is long – the data is available between April 1989 and September 2012 – and it will used to apply the statistical procedures.

4.1. Heteroscedasticity and trend modelling

Generally, environmental time series can be non-stationary in a variety of ways. The most common case occurs when data present a trend and/or heteroscedasticity. When data has variance in terms of time, i.e., $var(Y_t)=\sigma_t^2$, the most commonly applied procedure is the application of data transformation.

In several areas there are models that accommodate heteroscedasticity within their structure. For example, in econometrics, ARCH models are commonly employed in the modelling of financial time series that exhibit time-varying volatility. When an autoregressive moving average model (ARMA) is assumed for the error variance, the model is a generalized autoregressive conditional heteroscedasticity [19] model.

Heteroscedasticity in an environmental time series is commonly treated using the Box-Cox transformation [20]

$$y_\lambda^* = \frac{y^\lambda - 1}{\lambda}$$

which is indexed by λ, where y_λ^* is the transformation of the original observation y. Note that for $\lambda=0$, the transformation is the natural logarithm of y. On the one hand, the Box-Cox transformation is not suitable for all heteroscedasticity types. On the other hand, this transformation, among others, changes the data magnitude hindering the interpretation of the model.

As will be shown later, this work considers a class of models (state space models) that is able to incorporate some types of heteroscedasticity and does not require data transformation. A time series frequently has a trend, i.e., it is not an observed process with a constant mean. In general, the trend is modelled using a deterministic function $f(t)$, usually a polynomial function, power function or logarithmic function. The adjustment of the function $f(t)$ to data is generally performed using the least square method:

$$\min \sum_t \left[y_t - f(t) \right]^2 .$$

Nevertheless, the most common function is of the type $f(t)=\alpha+\beta t$. However, the simplicity of this function is not always appropriate to model real data, see for instance Figure 10. The DO concentration data for the Carvoeiro water-monitoring site represented in Figure 10 corresponds to an extended period of time. The graphical representation clearly indicates that the time series is not stationary in mean and the variability does not follow a monotonic function in time. Furthermore, there is no unique function among those referred above that globally fit the time series in order to model the trend. Thus, we can identify a trend function for the time series defined by sub-periods of time.

Figure 10 suggests that there is a time where the trend function changes its expression, possibly with a non-linear function prior to that time and a linear form following it. For simplicity and according to the exploratory analysis, we considered the most suitable function to the first part of the series to be a quadratic function and for the second part, a constant.

After fitting a regression model with a quadratic trend up to a certain time t_0, to all possible times t_0 and a constant after that instant, we chose the value of t_0 with the smallest residual sum of squares [21]. This procedure led to $t_0 = 139$, i.e., October 2000 and the estimated trend was:

$$\hat{T}_t = \begin{cases} -0.0014t^2 + 0.01986t + 6.8991 & \text{if} \quad t \le 139 \\ 9.128 & \text{if} \quad t > 139. \end{cases}$$

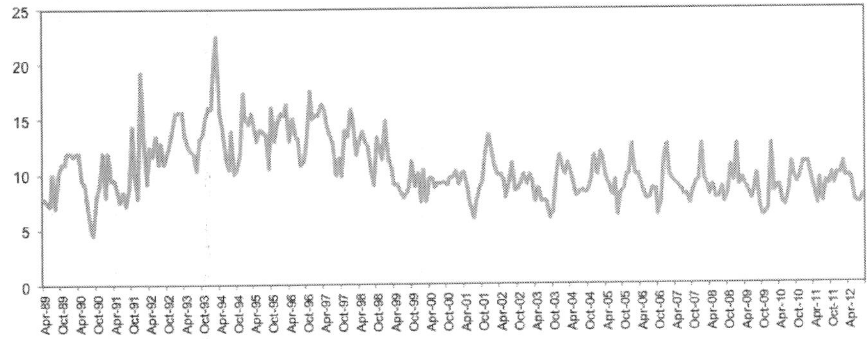

Figure 10. Monthly DO concentration in the Carvoeiro water-monitoring site from April 1989 to September 2012.

Figure 11 presents the series resulting from the subtraction of the adjusted trend to the original data, $Y_t - T_t$. As is shown, this new series does not present a deterministic trend; however, as is the norm for monthly environmental data, seasonal behaviour is present.

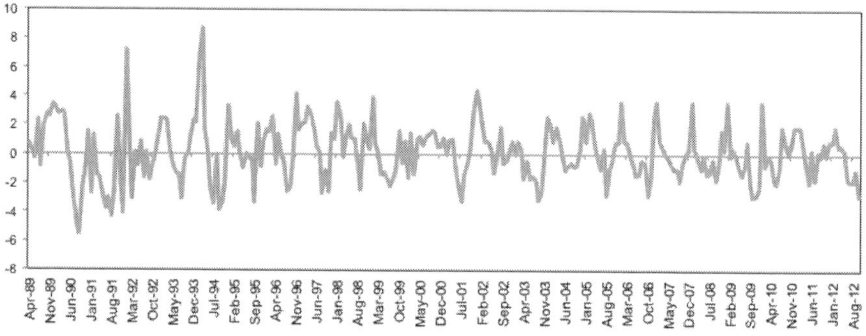

Figure 11. Residual series from the trend adjustment to the Carvoeiro DO concentration time series.

The easiest way to handle seasonality is to subtract from January's data the overall January average, from February's data the overall February average, etc. [22]. An alternative method involves estimating both trend and seasonality coefficients together. In this case, it is necessary to define a set of dummy variables as,

$$x_{ti} = \begin{cases} 1 & \text{if } t = (j-1)P + i, \text{for } some\ j = 1, \ldots, n \\ 0, & \text{otherwise} \end{cases}$$

where P is the period (in our case $P = 12$). If the trend component is polynomial, as previously considered, the model shows unique representation through a linear regression model. As it is generally appropriate that the sum of the seasonal coefficients must be zero, the linear model should be re-parameterized in order to overcome the linear dependence of variables x_{ti}. Thus, the model can be presented as:

$$Y_t = \beta_1 t + \beta_2 t^2 + \ldots + \beta_q t^q + \sum_{i=1}^{p} x_{ti} s_i + \varepsilon_t$$

where ε_t is white noise. The vector of coefficients $(\beta_1, \beta_2, \ldots, \beta_q, s_1, s_2, \ldots, s_P)$ is estimated by the least squares method. In order to facilitate the interpretation of results, we can consider the model with a constant parameter β_0, which leads to the following transformations:

$$\beta_0 = \bar{s}\ and\ s_i^* = s_i - \bar{s}, i = 1,2,\ldots,P$$

In general, these two approaches produce similar estimates. The first method is easier to implement and less laborious than the second. Furthermore, the regression model provides optimal estimates in certain conditions that assume that the errors ε_t are not time correlated, an assumption that is not valid for the data in the current analysis. Hence, for simplicity and considering the data characteristics, we adopted the first method, that is, the decomposition procedure that estimates trends and seasonal separately. Table 2 presents the estimates for the seasonal coefficients of the Carvoeiro data after the subtraction of the adjusted trend component.

Month	Seasonal coefficient
Jan	2.23
Feb	1.50
Mar	1.03
Apr	0.31
May	-0.24
Jun	-0.92
Jul	-1.01
Aug	-1.05
Sep	-1.40
Oct	-0.41
Nov	0.31
Dec	1.24

Table 2. Estimates for seasonal coefficients.

As expected, from the DO perspective, the water quality was better in the winter months and worse during the summer months. The DO concentration is associated with weather conditions, particularly with precipitation amounts [23].

The Box-Cox transformation is associated, for example, with multiplicative models where the time series components are multiplicative as $Y_t = T_t \cdot S_t \cdot \varepsilon_t$. However, there are also other types of heterocedasticity. Figure 12 shows for each month both the overall monthly averages and standard deviations of the DO concentration in Carvoeiro. This analysis indicated that variability had not been constant throughout the year. Indeed, the linear correlation coefficient between averages and standard deviations was 0.44, which showed that when the monthly average was large, the standard deviation tended to be large as well. This type of heteroscedasticity will be taken into account in the modelling procedure via the state space model approach.

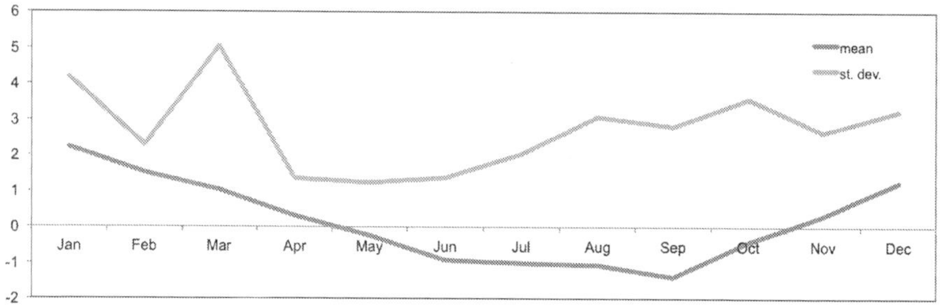

Figure 12. Monthly averages and standard deviation of DO concentration, subtracted according to trend, in the Carvoeiro water-monitoring site from April 1989 to September 2012.

4.2. Temporal dependence

It is well known that there exists a certain level of persistence in the behaviour of nature [22]. Indeed, environmental data is influenced by meteorological conditions that may persist from one month to another [21]. The series of residuals pertaining to the DO concentration in Carvoeiro does not have the characteristics of white noise. Figure 12 shows that there is a significant temporal autocorrelation according to an autoregressive process of order 1, an AR(1). Thus, stochastic modelling of residuals is needed in order to accommodate the serial correlation.

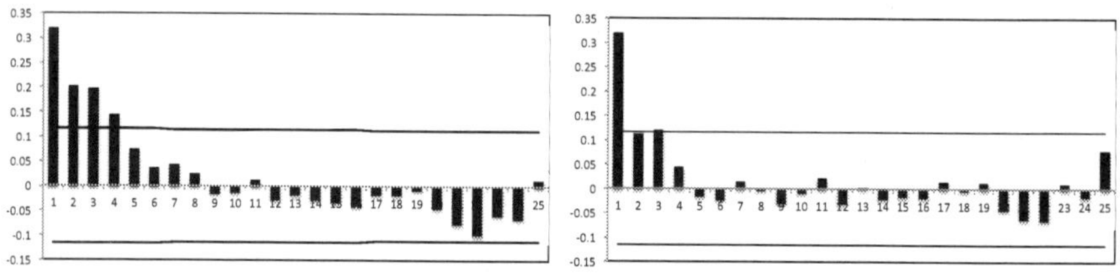

Figure 13. Sample autocorrelation function (ACF – left) and sample partial autocorrelation function (PACF – right) of the residuals.

When modelling a time series using the decomposition method approach, the procedure that is generally adopted for finding the best fit of an ARIMA model to past values of a time series is the Box-Jenkins methodology (for more details, see [24]. The modelling of the dependence of environmental data significantly influences decisions in the application of other modelling statistical procedures such as, for example, the existence of non-stationarities in a series or in change point detection procedures. Therefore, in section 5, we consider a class of models that allows for incorporating several components as the temporal dependence in a dynamic manner and which will be used to model the DO concentration in the water-monitoring of Carvoeiro.

5. Time series modelling

State space models are a class of extremely versatile time series models. Their popularity is mainly due to the Kalman filter. This algorithm, [25], has been used in many different areas to describe dynamic systems evolution. The primary goal of the Kalman filter algorithm is to find estimates of unobservable variables based on observable variables related through a set of equations designated by a state space model (SSM).

In general, such models are defined by the following equations:

$$Y_t = H_t \beta_t + e_t$$
$$\beta_t = \Phi \beta_{t-1} + \varepsilon_t$$

The first equation is the measurement equation and relates the $n \times 1$ vector of observable variables, Y_t, with the $m \times 1$ vector of unobservable variables, β_t, referred to as *states*. The $n \times m$ matrix H_t is a matrix of known coefficients and e_t is a white noise $n \times 1$ vector, called the measurement error, with covariance matrix $E(e_t e_t^{'}) = \Sigma_e$.

Furthermore, the vector of states β_t varies in time according to the second equation, the state equation, where Φ is a $m \times m$ matrix of autoregressive coefficients and ε_t is a white noise $m \times 1$ vector with covariance matrix $E(\varepsilon_t \varepsilon_t^{'}) = \Sigma_\varepsilon$. The disturbances e_t and ε_t are assumed to be uncorrelated, that is, $E(e_t \varepsilon_s^{'}) = 0$ for all t and s.

A subclass of these models with a particular interest arises when the state vector is a stationary process with mean μ, $E(\beta_t) = \mu$. In this case, to ensure that the state process and the state equation follows a stationary VAR(1) with a mean μ, it is assumed that the eigenvalues of the autoregressive matrix Φ are inside the unit circle. For more details, see [26].

5.1. A brief description of the Kalman filter algorithm

As the state β_t is unobservable, it is necessary to obtain its predictions. The Kalman filter algorithm uses an orthogonal projection of the state based on the available information up to the moment. Assuming that the parameters of a state space models are known, the Kalman filter recursions provide the best linear predictors for filtering and forecasting the vector of states.

Let $\hat{\beta}_{t|t-1}$ represent the predictor of β_t based on the information up to time $t-1$ and $P_{t|t-1}$ be its mean square error (MSE). As the orthogonal projection is a linear estimator, the predictor for the observable variable, Y_t, is given by

$$\hat{Y}_t = H_t \hat{\beta}_{t|t-1}.$$

When at time t, Y_t is available, the prediction error or *innovation*, $\eta_t = Y_t - \hat{Y}_t$, is used to update the prediction of β_t, the filter prediction, through the equation

$$\hat{\beta}_{t|t} = \hat{\beta}_{t|t-1} + K_t \eta_t$$

where $K_t = P_{t|t-1} H_t (H_t P_{t|t-1} H_t' + \Sigma_e)^{-1}$ is called the Kalman gain matrix. Furthermore, the MSE of the update predictor $\hat{\beta}_{t|t}$ verifies the relation $P_{t|t} = P_{t|t-1} - K_t H_t P_{t|t-1}$. In turn, at time t, the forecast for the state vector β_{t+1} is given by the equation

$$\hat{\beta}_{t+1|t} = \Phi \hat{\beta}_{t|t}$$

with MSE matrix $P_{t+1|t} = \Phi P_{t|t} \Phi' + \Sigma_\varepsilon$.

When the disturbances are Gaussian, the Kalman filter provides the minimum mean square estimate for β_t.

5.2. Parameter estimation

The application of the Kalman filter to predict the unknown variables requires the estimation of the model's parameters.

The vector of parameters $\Theta = \{\Phi, \Sigma_e, \Sigma_\varepsilon\}$ and the mean vector μ if the state is stationary must be estimated from the data. When the Gaussian distribution is suitable to the disturbances e_t and ε_t the conditional log-likelihood of a sample $Y_1, Y_2, \ldots, Y_n$ is given by:

$$\begin{aligned} \ln L(\Theta; Y_1, Y_2, \ldots, Y_n) &= \sum_{t=1}^{n} \ln f_{\hat{Y}_{t|t-1}}(Y_t \mid Y_1, Y_2, \ldots, Y_{t-1}) \\ &= -\frac{n}{2}\ln(2\pi) - \frac{1}{2}\sum_{t=1}^{n}\ln(|\Omega_t|) - \frac{1}{2}\sum_{t=1}^{n}\eta_t' \Omega_t^{-1} \eta_t \end{aligned}$$

where $\Omega_t = H_t P_{t|t-1} H_t' + \Sigma_e$. The maximum likelihood estimates are obtained by maximizing the log-likelihood, i.e.,

$$\hat{\Theta}_{ML} = \arg\max \ln L(\Theta; Y_1, Y_2, \ldots, Y_n).$$

The state space model's parameters are estimated by the maximum likelihood via numerical procedures.,The Newton-Raphson method can be adopted or, more often, it is employed the EM algorithm [24].

If the Gaussian distribution is not appropriate or the optimal properties are not required in the modelling procedure, other estimators can be considered. For example, distribution-free estimators can be adopted, which do not assume any distribution for the disturbances (see [27, 23]).

5.3. A mixed-effect state space model

In order to incorporate the temporal correlation and to accommodate the heterogeneity of the variances of the sub-series of each month in terms of DO concentration in Carvoeiro, we adopted a mixed-effect state-space model. The model is defined as:

$$\begin{aligned} Y_t &= (T_t + S_t)\beta_t + e_t \\ \beta_t &= \mu + \phi(\beta_{t-1} - \mu) + \varepsilon_t \end{aligned}$$

where T_t is the trend component, S_t is a periodic function with 12 averages as previously mentioned, the process $\{\beta_t\}$ is a stationary AR(1) with mean μ and Gaussian errors ε_t and e_t is the measurement error, assumed to be a Gaussian white noise process.

The model has two primary components: a regression structure, which incorporates both the trend and the seasonality, and an unobservable process $\{\beta_t\}$, the state. It is assumed that the state process will calibrate the deterministic structure $T_t + S_t$ previously estimated by $\hat{T}_t$ and $\hat{S}_t$ in presented in Table 2.

Another important feature of the model is that by its own formulation, it allows the existence of heterogeneity of variances previously identified. That is, the stochastic calibrator factor allows for the dynamic modelling of the heteroscedasticity during the year and incorporates the time dependence identified through the ACF and PACF functions in Figure 12.

Table 3 presents the maximum likelihood estimates of $\Theta = \{\mu, \phi, \sigma_\varepsilon^2, \sigma_e^2\}$ and their respective standard errors. As expected, the mean of the calibration factor was close to 1. This meant that the deterministic component $T_t + S_t$ represented the global behaviour of data. On the other hand, the calibration factor followed a stationary autoregressive process AR(1) process since $|\hat{\phi}| < 1$.

$\hat{\mu}$	$\hat{\phi}$	$\hat{\sigma}_\varepsilon^2$	$\hat{\sigma}_e^2$
1.00077	0.43001	0.01790	0.24292
(0.00085)	(0.00503)	(0.00024)	(0.01854)

Table 3. Maximum likelihood estimates of the mixed-effect state space model (standard errors in brackets).

Once the model had been well identified, the Kalman filter was run in order to obtain the forecasts and filtered predictions of the states β_t. The Kalman filter provided predictions to the states and their respective MSE estimates. Figure 13 represents the filtered predictions $\hat{\beta}_{t|t}$ of the calibration factors for each month and their respective empirical confidence intervals at a 95% level obtained by:

$$\beta_t = \hat{\beta}_{t|t} \pm 1.96\sqrt{\hat{P}_{t|t}}.$$

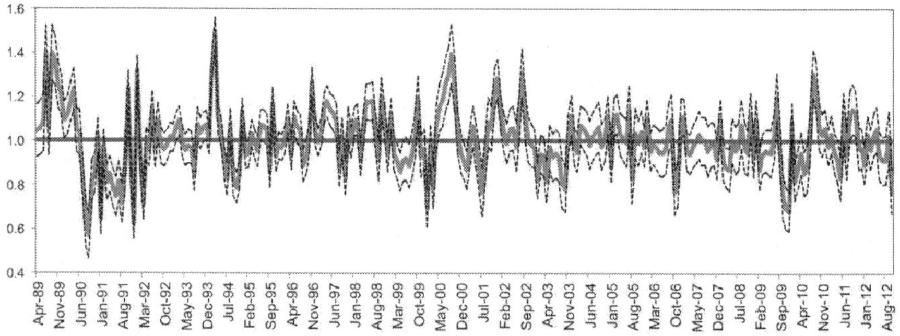

Figure 14. Representation of filtered predictions $\hat{\beta}_{t|t}$ and their respective empirical confidence intervals.

In order to assess the adjustment of the model, one step-ahead forecasts were computed with respective confidence intervals at 95% (see Figure 14):

$$Y_t = \hat{Y}_{t|t-1} \pm 1.96\sqrt{\widehat{MSE}_{t|t-1}}.$$

where $\widehat{MSE}_{t|t-1} = (\hat{T}_t + \hat{S}_t)^2 \hat{P}_{t|t-1} + \hat{\sigma}_e^2$.

The percentage of observations outside of the respective empirical confidence interval was 6.36%. The residuals analysis showed that there were some outliers; the biggest residual occurred in January 1992.

Normality was rejected at a 5% significant level, considering the Kolmogorov-Smirnov test, since the p-value was 2.8%. However, the Gaussian distribution was not rejected (K-S p-value=20.0%) when the largest outliers were not considered. As state space models are associated to the Kalman filter, in general, the impact of the outliers turns out not to be significant, since they are attenuated through the filtering procedure.

As can be seen in Figure 14, a large proportion of the observations outside the respective empirical confidence interval belonged to the first years of the series. This may indicate the existence of different structures in the observable variables, which can be found through the analysis of change points in the state variables. This issue is discussed in the following section.

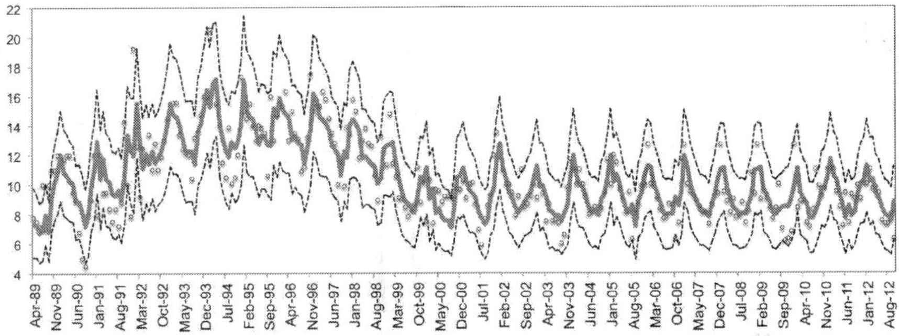

Figure 15. Dissolved oxygen concentration and the step-ahead forecast with respective confidence intervals at 95%.

5.4. Change point detection

The detection of structural changes in environmental data is one of the most interest issues in the applied statistical analysis. In the past decade, there has been significant progress within change point detection. However, these methodologies still require more complex statistical procedures or computational requirements. For example, in [28], a simple BIC-like multiple change point penalty is proposed that is based on the total number of change points. For further recent developments regarding change point detection in environmental applications, [29,30].

The state space modelling approach adopted in this work allows for applying standard change point detection procedures, since this model incorporates several data characteristics. Thus, as these properties of environmental data are taking into account in the modelling procedure, simple change-point detection techniques can be applied if adequately chosen.

This subsection deals with change point detection in the state process instead of the observed process Y_t. This detection is made through the filtered predictions of the calibration factors, $\hat{\beta}_{t|t}$, obtained in the filter procedure. It is reasonable to investigate changes in the structure of β_t that may be relevant to the water-monitoring process. For this purpose, the filtered predictions of the state process can be analysed in relation to the existence of one change (or more) in its mean as the best prediction of an AR(1) process.

To do so, we used maximum type test statistics, referred in [31], alongside their correct adaptation to an autoregressive process. The basic test evaluates the existence of a change point in the mean of independent and identically distributed Gaussian variables in an unknown time.

The hypotheses of the basic test are:

$H_0: \beta_1, \beta_2, \ldots, \beta_n$ are independent and distributed according to the same Gaussian distribution $N(\mu, \sigma^2)$

vs.

$H_0: \exists k \in \{1, \ldots, n-1\}: \beta_1, \ldots, \beta_k$ are distributed according to $N(\mu_1, \sigma^2)$ and $\beta_{k+1}, \ldots, \beta_n$ are distributed according to $N(\mu_2, \sigma^2)$, with $\mu_1 \neq \mu_2$.

The test for the existence of the change point k can be performed by a statistic test known as the maximum type. Since the variance is generally unknown, this approach consists of a sequence of t tests and the analysis of the significance of their maximums. Thus, the test statistic is given by:

$$T_n = \max_{1 \le k < n} | T_k | = \max_{1 \le k < n} \sqrt{\frac{(n-k)k}{n}} \frac{\left| \overline{\beta}_{k|k} - \overline{\beta}^*_{k|k} \right|}{s_k}$$

where

$$\overline{\beta}_{k|k} = \frac{1}{k} \sum_{i=1}^{k} \hat{\beta}_{i|i}, \overline{\beta}^*_{k|k} = \frac{1}{n-k} \sum_{i=k+1}^{n} \hat{\beta}_{i|i} \text{and } s_k = \sqrt{\frac{1}{n-2} \left[\sum_{i=1}^{k} \left(\hat{\beta}_{i|i} - \overline{\beta}_{k|k} \right)^2 + \sum_{i=k+1}^{n} \left(\hat{\beta}_{i|i} - \overline{\beta}^*_{k|k} \right)^2 \right]}.$$

The null hypothesis is rejected if the statistics T_n are larger than a corresponding critical value. However, the calculation of the exact critical value is not easy, due to the test statistic being the maximum of the sequence $T_1, T_2, \ldots, T_{n-1}$. Nevertheless, approximated critical values can be obtained by different methods, namely:

i. the Bonferroni inequality and its improvement;

ii. asymptotic distribution;

iii. through simulation.

Table 4 replicates the approximated critical values obtained through simulation and presented in [22]. However, these critical values assume that the observations are uncorrelated and according to [32], it is necessary to correct them. Thus, in order to obtain the corrected critical value that takes into account the time correlation of an AR(1), the initial critical value must be multiplied by $[(1+\hat{\phi})/(1-\hat{\phi})]^{1/2}$.

The simplest method for detecting multiple change points is through binary segmentation.

The first work to propose binary segmentation within a stochastic process setting was [33]. Binary segmentation is a generic technique for multiple change-point detection in which, initially, the entire dataset is searched for one change-point, typically using a CUSUM-like procedure. When a change-point is detected, the data are split into two (hence, 'binary') subsegments, defined by the detected change-point. A similar search is then performed on each sub-segment, possibly resulting in further splits. The recursion on a given segment continues until a certain criterion is satisfied [34].

n	5% critical value	1% critical value
50	3.15	3.76
100	3.16	3.71
200	3.19	3.72
300	3.21	3.73
500	3.24	3.73

Table 4. Approximated critical values obtained by simulation.

In order to apply the change point detection procedure to the state filtered prediction of Carvoeiro, it was necessary to interpolate the critical value at 5% to n=283, which is 3.207. Taking into account the autoregressive parameter estimate presented in Table 3, the corrected critical value at 5% was noted as 5.079. Figure 15 represents the observed values of the statistics T_k, $1 \le k < 283$, as well as both corrected and uncorrected critical values at 5% in the first step of the binary segmentation procedure. As can be seen, the null hypothesis is rejected, i.e., a statistically significant change point that can be identified as April 1990.

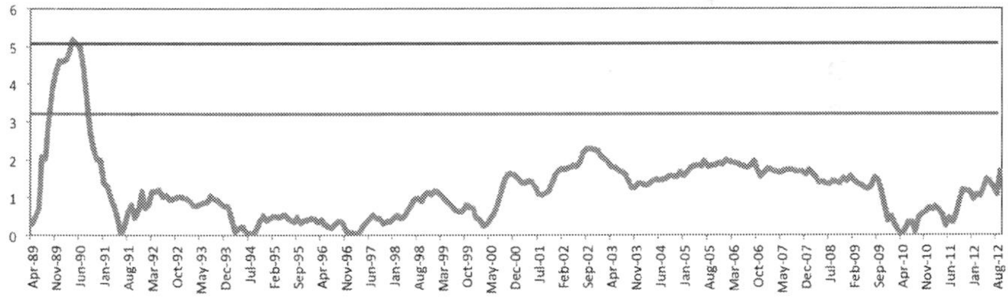

Figure 16. Observed values of the statistic T_k, $1 \le k < 283$; horizontal lines represent both critical values at 5% corrected and uncorrected to an AR(1).

Table 5 presents the results from the complete binary segmentation procedure, which ended when the observed value of the test statistics was less than the respective corrected critical value (the critical value was obtained by interpolation according to the sub-sample size). This procedure identified two change points: April 1990 and December 1991. Note that in the final step of the binary segmentation procedure, the largest value of the statistic tests was 3.202, which would lead to the detection of another change point if the correction to the AR(1) process had not been considered, since the uncorrected critical value was 3.200. However, considering the AR(1) correction, the critical value was 5.069, i.e., the observed statistic test was not statistically significant.

n_i	$T(n_i)_{obs}$	Month/Year	Critical value	Critical value
			uncorrected	corrected
283	5.1893	April 1990	3.2066	5.0790
270	5.7600	December 1991	3.2040	5.0749

Table 5. Results from the binary segmentation procedure.

The complete binary segmentation procedure identified two significant change points that corresponded to three different levels for the mean of the process $\{\beta_t\}$. In chronological order, the mean levels of the process $\{\beta_t\}$ were 1.190, 0.835 and 1.005. This means that during the first period of time, the DO variable was on average 19% higher than those predicted by the regression component. In the second period of time, the observations were on average 16.5% lower, while in the last period of time, observations were on average close to the regression model's predictions.

The identification of statistically significant change points provides information that can be useful to environmental investigators or technicians. Nevertheless, from a statistical point of view, this information allows for obtaining more accurate models with better adjustments. Thus, we must consider the state space model defined in subsection 5.3, but only where the state process $\{\beta_t\}$ has three average levels, one for each sub-series identified in the change point procedure.

Thus, taking three means μ_1=1.190, μ_2=0.835 and μ_3=1.005 according to the identified change points of April 1990 and December 1991, the percentage of observations outside of the respective empirical confidence interval, at 95% level, for the forecasted one step-ahead of the DO concentration in the new model is 5.30% instead of 6.36% in the first model. On the other hand, the final model exhibited a determination coefficient equal to R^2=0.711 instead of R^2=0.689 in the first model.

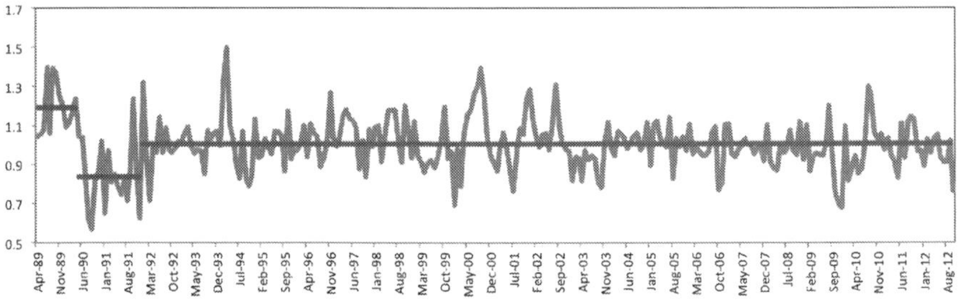

Figure 17. Representation of change points and the three mean levels of the state process.

6. Conclusions

The statistical analysis of water quality data is a relevant tool for the monitoring of ecosystems and water systems for human consumption. The information contained in water variable time series allows studying the past but also enables monitoring the present and planning the future. Biochemical analyses of water must be accompanied by statistical studies in order to identify patterns and analyse the evolution of water quality in its different forms.

This work presented a characterization of the River Vouga Basin in Portugal and focused primarily on the dissolved oxygen concentration variable. The complexity of databases renders more difficult to perform consistent statistical analyses of water quality. Statistical issues were presented and discussed in this paper to make their implementation easier for other researchers. Several problems were discussed and some solutions were provided, always keeping simplicity and practical relevance in mind.

Acknowledgements

Authors were partially supported by Portuguese funds through CIDMA – Center for Research and Development in Mathematics and Applications and the Portuguese Foundation for Science and Technology (FCT – Fundação para a Ciência e a Tecnologia") within project PEst-OE/MAT/UI04106/2013.

Author details

Marco André da Silva Costa* and Magda Sofia Valério Monteiro

*Address all correspondence to: marco@ua.pt

School of Technology and Management of Águeda, Center for Research & Development in Mathematics and Applications, University of Aveiro, Portugal

References

[1] Lopes JF. Silva CI. Cardoso AC. Validation of a water quality model for the Ria de Aveiro lagoon. Portugal. Environmental Modelling & Software 2008;23 479-494.

[2] AEA - Agência Europeia do Ambiente. Os recursos hídricos da Europa: Uma avaliação baseada em indicadores (Síntese). Copenhaga. Serviço das Publicações Oficiais da União Europeia; 2003.

[3] Solak CN. A̍cs E. Water Quality Monitoring in European and Turkish Rivers Using Diatoms. Turkish Journal of Fisheries and Aquatic Sciences 2011;11 329-337.

[4] Andersen MS. Effectiveness of urban wastewater treatment policies in selected countries: an EEA pilot study. European Environment Agency EEA Report n. 2/ 2005. Office for Official Publications of the European Communities. Luxembourg. 2005.

[5] Machado A. Silva M. Valentim H. Contribute for the evaluation of water bodies status in Northern Region. Revista Recursos Hídricos 2010;31(1) 57-63.

[6] Costa M. Gonçalves AM. Combining Statistical Methodologies in Water Quality Monitoring in a Hydrological Basin – Space and Time Approaches. In: Voudouris K. and Voutsa D. (ed.) Water Quality Monitoring and Assessment. Croatia: InTech; 2012. p121-142.

[7] Henriques V. Monitorização da qualidade da água na bacia Henriques hidrográfica do Vouga. Master thesis. Universidade de Aveiro; 2010.

[8] Portal da Água. Instituto da Água. I.P. (INAG). http://portaldaagua.inag.pt (accessed 10 July 2014).

[9] Sistema Nacional de Informação de Recursos Hídricos (SNIRH). Instituto da Água. I.P. (INAG). http://snirh.pt (accessed 10 July 2014).

[10] INE – Instituto Nacional de Estatística. http://www.ine.pt

[11] Lopes JF. Silva CI. Temporal and spatial distribution of dissolved oxygen in the Ria de Aveiro lagoon. Ecological Modelling 2006;197 67-88.

[12] Ahmad I. Mohmood I. Coelho J.P. Pacheco M. Santos M.A. Duarte A.C. Pereira E. Role of non-enzymatic antioxidants on the bivalves' adaptation to environmental mercury: Organ-specificities and age effect in Scrobicularia plana inhabiting a contaminated lagoon [ria]. Environmental Pollution 2012;163 218-225.

[13] Serodio J. Cartaxana P. Coelho H. Vieira S. Effects of chlorophyll fluorescence on the estimation of microphytobenthos biomass using spectral reflectance indices. Remote Sensing of Environment 2009; 113 1760-1768.

[14] Grupo uariadeaveiro. Universidade de Aveiro. http://www.ua.pt/riadeaveiro/

[15] http://maretec.mohid.com/Estuarios/MenusEstuarios/Descri%C3%A7%C3%A3o/descricao_RiaAveiro.htm (accessed 12 August 2014)

[16] Lopes A. - In: "Aveiro e o seu distrito ", n.º 23/25, 1977/78, pp. 9-13.

[17] Booty WG. David CL. Freshwater ecosystem water quality modelling. In: Davies. A.M. (ed.) Modelling Marine Systems. vol. II. Boca Raton: CRC Press; 1989. p387-431.

[18] Shifflett DS. Water and Sustainability. http://www.unc.edu/~shashi/TablePages/dissolvedoxygen.html (accessed 16 July 2014)

[19] Bollerslev T. Generalized Autoregressive Conditional Heteroskedasticity. Journal of Econometrics 1986;31 307–327.

[20] Box GEP. Cox DR. An analysis of transformations. Journal of the Royal Statistical Society Series B 1964;26 211-252.

[21] Alpuim T. El-Shaarawi A. Modeling monthly temperature data in Lisbon and Prague. Environmetrics 2009;20 835-852.

[22] Jarušková D. Some problems with applications of change-point detection methods to environmental data. Environmetrics 1997;8 469-483.

[23] Gonçalves A. Costa M. Predicting seasonal and hydro-meteorological impact in environmental variables modelling via Kalman filtering. Stochastic Environmental Research and Risk Assessment 2013; 27 1021-1038.

[24] Shumway RH, Stoffer D. Time series analysis and its applications: with R examples. New York, Springer; 2006

[25] Kalman RE. A new approach to linear filtering and prediction problems. Transactions of the ASME, Journal of Basic Engineering 1960; 82 (Series D) 35-45.

[26] Harvey A.C. Forecasting structural time series models and the Kalman filter. Cambridge, Cambridge University Press; 1996

[27] Costa M. Alpuim T. Parameter estimation of state space models for univariate observations 2010; 140(7) 1889–1902

[28] Caussinus H. Mestre O. Detection and correction of artificial shifts in climate series. Journal of the Royal Statistical Society 2004; 53 (Series C) 405-425

[29] Lu Q. Lund R.B. Lee T.C.M. An MDL Approach to the Climate Segmentation Problem. Annals of Applied Statistics 2010; 4 299-319.

[30] Li S. Lund R.B. Multiple Changepoint Detection via Genetic Algorithms. Journal of Climate 2012; 25 674-686.

[31] Costa M. Goncalves A. Application of Change-Point Detection to a Structural Component of Water Quality Variables. In Theodore et al. (ed.) Numerical Analysis and Applied Mathematics ICNAAM 2011, AIP Conference Proceedings 1389, American Institute of Physics; 2011 New York, p. 1565-1568.

[32] Jaruskova D. Change-point detection meteorological measurement. Monthly Weather Review 1996; 124 1535-1543

[33] Vostrikova L. Detecting 'disorder' in multidimensional random processes. Soviet Mathematics Doklady 1981;24 55-59.

[34] Fryzlewicz P. Wild binary segmentation for multiple change-point detection. Preprint, London School of Economics, http://stats.lse.ac.uk/fryzlewicz/wbs/ (accessed 12 August 2014).

Section 2

Methods and Practices in Determining and Improving Water Quality

Chapter 7

Comparative Assessment of Groundwater Quality in Rural and Urban Areas of Nigeria

A.M. Taiwo, A.T. Towolawi, A.A. Olanigan,
O.O. Olujimi and T.A. Arowolo

Additional information is available at the end of the chapter

http://dx.doi.org/10.5772/59669

1. Introduction

Groundwater resource is a significant source for the provision of good quality drinking water to humans and animals. Access to potable water is one of the Millennial Development Goals (MDGs) in all developing nations of the world. This goal is yet to be achieved because an estimated two billion people worldwide lack access to potable water [1]. In Nigeria alone, about 52% of the population lack access to safe drinking water [2]. Groundwater quality differs from place to place and this may therefore affect its suitability for consumption [3]. For example, land-use has been found to affect the quality of groundwater [2]. Polluted groundwater resource may initiate water-borne diseases such as gastroenteritis, cholera, typhoid fever and giardiasis [4]. However, when groundwater is adequately protected and well managed, it could be a good source of drinkable water. In some parts of India, groundwater accounts for the supply of drinking water to about 88% of rural population where access to basic infrastructures for water treatment is lacking [5]. In Nigeria, groundwater does not only serve the need of rural communities only but also provide drinking water for urban dwellers where there is epileptic supply of pipe-borne water [6].

The major threats to groundwater quality include domestic, commercial and industrial wastes and increased agricultural use of fertilizers and pesticides [7]. These pollutants may infiltrate into aquifers through seepage and thereby polluting it. In the recent times, nitrate concentration of groundwater resources in most parts of Nigeria has increased due to rise in the construction of soak-away septic tanks in individual houses all over the country [6, 8]. Nitrate is dangerous to infants less than six months old. It causes a disease known as methemoglobinemia [9]. In the coastal areas, intrusion of saline water may be problematic to groundwater resource [8]. Microbiological pathogens can also pose a significant danger to consumption of groundwater. Nigeria has suffered some worst cases of cholera epidemic in recent times. In November 2001, a total of 724 cases and 52 deaths from cholera outbreak were reported in three communities in Nigeria (Kano, Kwara and Akwa Ibom states) [10]. Similarly, in December 2001, a total of 2050 cases and 80 deaths of cholera epidemic were recorded in Kano Metropolis [10]. In the developing countries, two million infants deaths are observed yearly due to consumption of unsafe drinking water [11]. This study therefore assessed the groundwater quality at both rural and urban areas in Nigeria in order to evaluate its suitability for consumption.

2. Material and methods

2.1. Geological description of the study areas

The two study areas can be found in the old Western Region of Nigeria. Figure 1 shows the geological map of Nigeria. An extensive discussion of the geology of Nigeria can be seen elsewhere in Jones and Hockey [12]. Two major geological formations characterises the south-western Nigeria. These include Basement Complex and Dahomey Basins. Under these two main formations are Abeokuta formation, Alluvium, Ewekoro formation, Ilaro formation and Coastal Plain Sands. According to Oyedele et al. [13], the south-western Nigeria's geology consists of rocks of older granite (pegmatite, granite-gneiss, grandiorite, migmatite and quartz diorite), charnockitic (pyroxene-diorite and metagabbro) and gneiss complex (quartzitite, biotite and biotite hornblende-gneiss mica-schist, amphibolite schist and granulitic gneiss).

2.2. Sampling and analysis

Groundwater were sampled from selected wells at rural (Forest Reserve near Benin) and urban (Ketu, Lagos) areas during rainy and dry seasons in 2011. Fifteen groundwater samples were sampled in the rural location while eight samples were collected in the urban location per season. Upon collection of the groundwater samples, parameters such as pH, temperature, colour, turbidity, total dissolved solids (TDS) and electrical conductivity (EC) were measured in situ using appropriate instruments (see Table 1).

Total suspended solids (TSS) were gravimetrically determined while total solids were calculated by summing TDS and TSS. Total hardness, alkalinity, acidity and chloride were measured titrimetrically and nitrate was determined by the sodium salicylate colorimetric method. Metals including Ca, Mg, Fe, Pb, Zn, Cu and Mn were measured using atomic absorption spectrometry (AAS). Prior to the determination of these metals, 100 mL of groundwater samples were digested with 10 mL concentrated nitric acid and heated for 30 minutes on a hot plate. After allowing to cool, the digested sample was transferred to a 100 mL volumetric flask and made to the mark. The digested samples were analysed with Atomic Absorption Spectrophotometer (Buck Scientific 200).

http://dx.doi.org/10.5772/59669

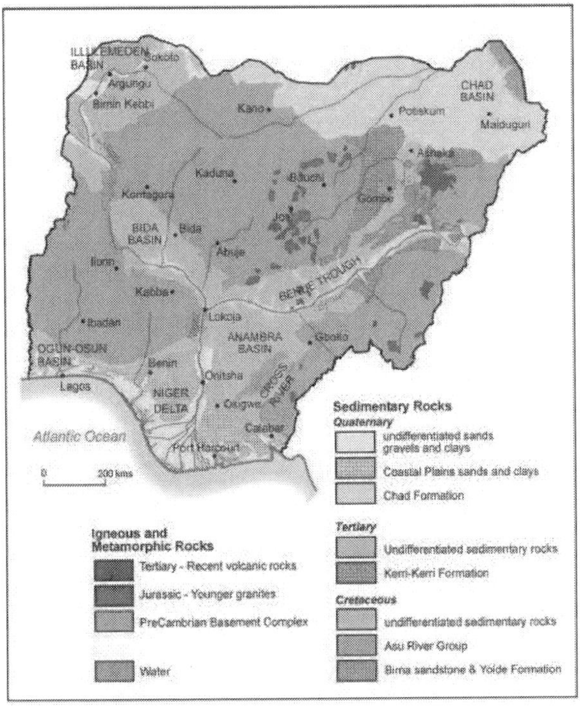

Figure 1. Geological map of Nigeria depicting Lagos and Benin where the groundwater wells were sampled [14].

Parameters	Methods of Analysis
pH/EC/TDS	Combined pH/EC/TDS meter (Combo HI 98130, Hanna, USA)
Temperature	Thermometer
Colour	Spectrometry
Turbidity	Turbidimetry
TSS	Gravimetry
Total Hardness	Titrimetry
Total Alkalinity	Titrimetry
Total Acidity	Titrimetry
Chloride	Silver nitrate [15]
Nitrate	Sodium Salicylate [16]
Metals	Atomic Absorption Spectrometry

Table 1. Analytical methods of analysis of groundwater samples

2.3. Statistical analysis

Data collected was analysed for descriptive statistics, analysis of variance and principal component analysis (PCA) using SPSS for Windows version 13.

3. Results and discussion

Figures 2-4 show the results (mean) of physical and chemical parameters measured in the sampled groundwater from the rural and urban sites. The results also indicated the seasonal variations at the two monitoring stations. The groundwater pH at both sites was generally less than 6.0 indicating slight acidic condition. This may pose a risk for consumption due to metal toxicity. Under a low pH condition, metals tend to go into solution thereby making it readily available for exposure [17, 18]. Jarup [19] highlighted deleterious effects of heavy metals contaminations. Most of the metals investigated in this study (except Fe) were observed to be below detection limits (0.01 mg/L for Pb, Zn, Mn and Cu of the Atomic Absorption Spectrophotometer. Temperature data of the groundwater samples were less than 30°C.

Turbidity mean values were 19 ± 31 and 29 ± 54 NTU at the urban area during wet and dry seasons, respectively. These turbidity values at the rural site were 6.8 ± 2.8 and 3.7 ± 3.2 NTU for wet and dry seasons, respectively. The groundwater samples were very turbid at the urban site with up to 280 and 484% rise above the WHO permissible value of 5.0 NTU in drinking water [20]. A slight elevated turbidity value (35% higher than the WHO standard) was also observed in rural groundwater samples during wet season. Water turbidity usually denotes the presence of suspended substances, and may interfere with disinfection as well as harbouring microorganisms in water [21]. Bacterial growth could also be stimulated in high turbid water [22].

The mean values of colour of the groundwater samples at the urban site were 105 ± 169 and 126 ± 229 TCU during wet and dry seasons respectively while 19.53 ± 7.3 and 11.6 ± 8.7 TCU were recorded during wet and dry seasons in rural location. These colour mean values were significantly higher than the WHO standard of 15.0 TCU in drinking water by 30-740% except that of rural dry season. The obtained colour concentration at the urban site was extremely high and indicated pollution. FAO [23] reported that elevated colour concentration might suggest coloured organic substances and the presence of metals such as Fe, Mn and Cu. Although Cu and Mn were observed to be below detection limit in this study; Fe concentration was observed to be significantly higher concentrations than the WHO standard in groundwater samples collected at the urban site. According to the WHO, drinking water with colour above 15 CTU may attract aesthetic displeasures. It has also been established that colour in water may modify the toxicity of metals [24]. The extreme values of colour observed in the urban groundwater therefore suggest groundwater pollution at these monitoring sites thereby rendering the water unsafe for drinking.

The EC values also followed similar pattern like those of TDS, TS, colour and turbidity with higher concentrations observed at the urban site; but the values were within the permissible

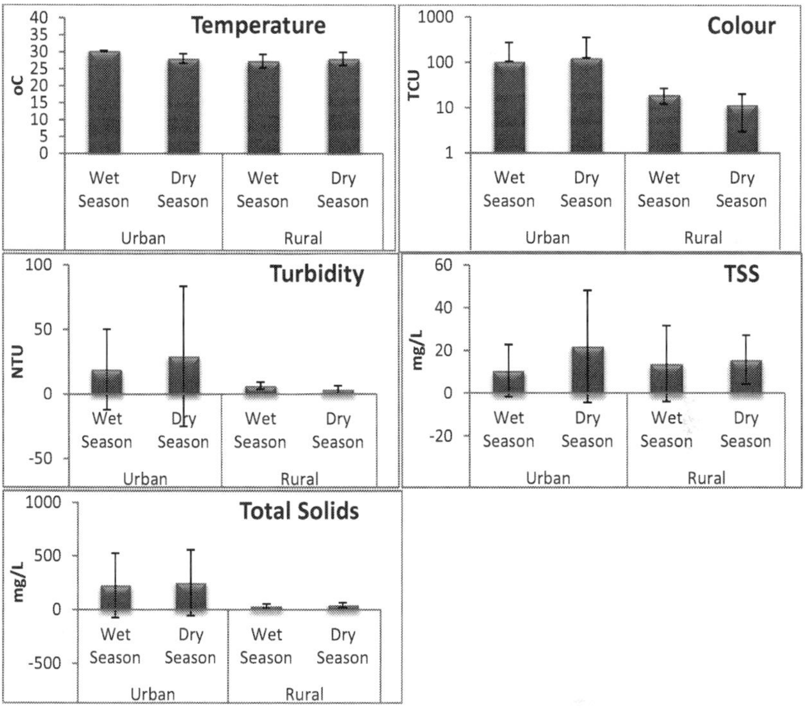

Figure 2. Comparison of the results of seasonal physical parameters of groundwater in urban and rural sites (Whiskers on the bars show the standard deviation values).

limit of 400-1250 μS/cm described by Mento [25]. Most of the studies carried out in Lagos have recorded elevated values of EC up to 43200 μS/cm [26]. A TSS mean value up to 22 ± 26 mg/L was observed during the dry season at the urban site (Figure 2). Since there's no permissible standard established for TSS in drinking water, it is difficult to ascertain the severity of the TSS values obtained in this study. For surface water protection, Roberts [27] has recommended a TSS value of 25-80 mg/L.TSS has been reported to harbour dangerous microorganisms [28]. The results of total hardness, alkalinity, acidity and chloride values (Figure 3) were generally high at the urban location; but were less than the WHO permissible standards. Nitrate data showed highest value at the rural site during the dry season (0.33 ± 0.41 mg/L); but below the WHO standard of 10 mg/L.

The level of iron in the groundwater samples at the urban site only was about 57 and 62 % higher than the WHO permissible level of 0.3 mg/L in drinking water [20] in samples collected in wet and dry seasons, respectively. Fe is an essential element in the body that forms the integral part of haemoglobin. However, at elevated concentration in the tissue, it can lead to conjunctivitis, choroiditis, and retinitis [29]. Elevated concentration of Fe in drinking water

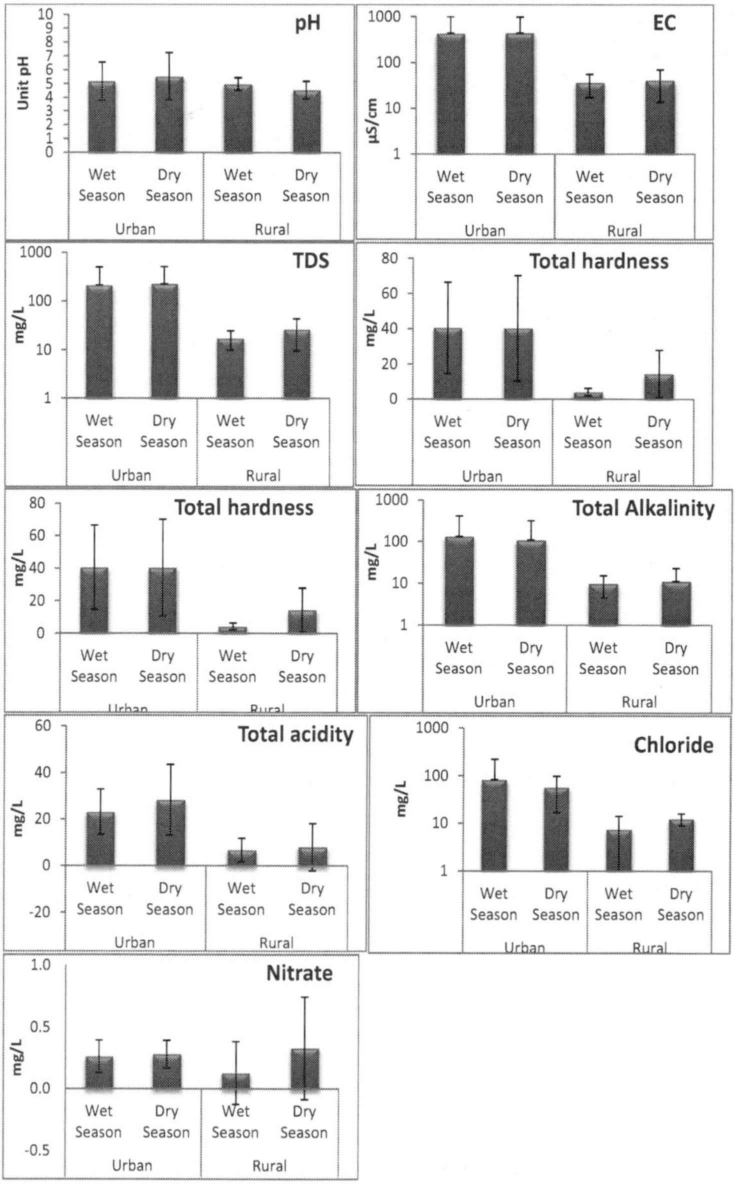

Figure 3. Comparison of the results of seasonal chemical parameters of groundwater in urban and rural sites (Whiskers on the bars show the standard deviation values).

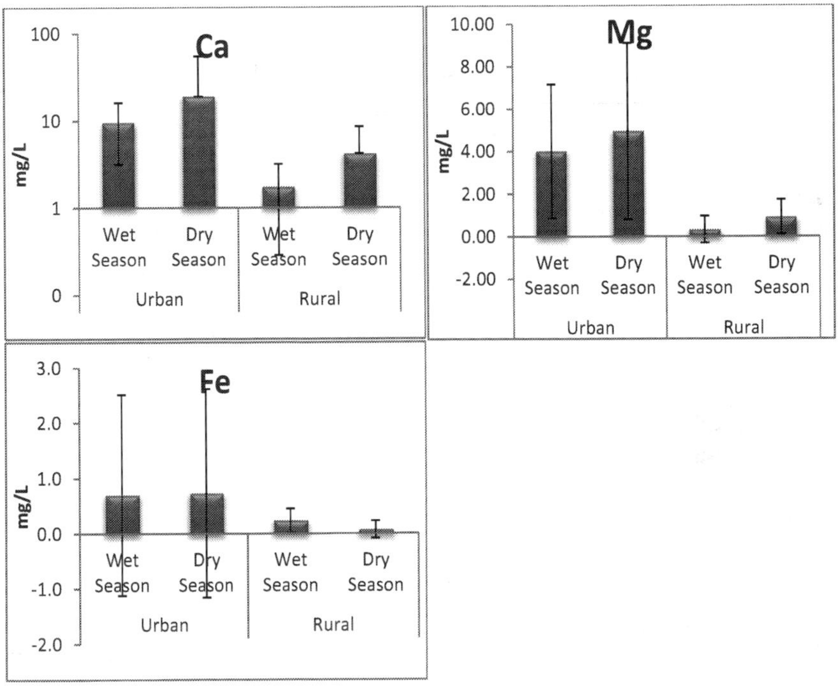

Figure 4. Comparison of seasonal levels of metals in groundwater in urban and rural sites (Whiskers on the bars show the standard deviation values)

may change its taste. The results obtained for Pb, Zn, Mn and Cu were below the detection limit.

Generally, most groundwater parameters (e.g. colour, turbidity, TSS, TS, TDS, EC, total alkalinity, total hardness, total acidity, Ca, Mg and Fe) showed a significant higher value at the urban site compared to the rural area (Table 2). This shows that the level of pollution of groundwater resources at the urban area is very high. However, the levels of industrial activities at the sampling locations in the urban site appeared to be minimal when this result is compared to that of Ayedun et al. [30] at an industrial site in Ifo. The results of this study is generally less than that of Adekunle et al. [31] for groundwater sampled in the rural area of Igbora, Oyo state, where the major contaminant of rural groundwater was organic waste [31].

Seasonal effects of groundwater quality were mostly observed for parameters such as turbidity, Ca and Mg at both sampling sites. There were instances where groundwater parameters were higher seasonally in one particular site. For instance, turbidity,TSS and total acidity were significantly higher during dry season in urban site while nitrate was higher during wet season in the rural site.

Parameter	F	Sig.	Parameter	F	Sig.
pH	3.560	.066	T-Hard	30.929	**.000**
Temp	7.028	**.011**	T-Alkal	6.212	**.017**
EC	16.296	**.000**	T-Acidity	36.552	**.000**
Color	7.945	**.007**	Cl	11.255	**.002**
Turb	5.832	**.020**	NO_3	.225	.637
TS	14.185	**.000**	Ca	5.995	**.018**
TDS	15.693	**.000**	Mg	32.315	**.000**
Fe	2.905	.095			

Temp-temperature, EC-electrical conductivity, Turb-turbidity, TDS-total dissolved solids, TSS-total suspended solids, TS-total solids, T-hard-total hardness, T-Alkal-total alkalinity, T-acidity-total acidity,

Table 2. ANOVA tests of groundwater parameters at the study areas

The rotated principal component analysis of groundwater samples collected at the urban site is presented in Table 3. The varimax PCA identified three factors with 86.6 % of the variance explained. Factor 1 which represents 58.4 % of the variance is characterized by significant loadings for EC, TS, TDS, TSS, total hardness, total alkalinity, Cl, NO_3, Ca and Mg. This factor might represent a mixed pollution sources from waste dump leachates, industrial effluents, agricultural run-off and seepage from soak-away or septic tanks [32]. Factor 2 (accountable for 20.6 % of the explained variance) is highly loaded for pH, colour, TSS, total acidity and Fe.

	Component			Communalities
	1	2	3	
pH	.524	**.712**	.075	**.787**
Temperature	.003	-.077	**-.811**	**.664**
EC	**.973**	.193	-.019	**.985**
Colour	.442	**.885**	.036	**.979**
Turbidity	.327	**.940**	.069	**.995**
TS	**.965**	.233	.031	**.986**
TDS	**.972**	.197	.014	**.984**
TSS	**.631**	**.654**	.254	**.891**
T-Hardness	**.866**	.310	.089	**.855**
T-Alkalinity	**.966**	.190	-.092	**.978**
T-Acidity	-.200	**.651**	.285	**.545**
Cl	**.828**	.041	-.425	**.868**
NO_3	**.733**	-.285	-.014	**.618**
Ca	**.705**	.245	**.490**	**.798**
Mg	**.967**	.063	.211	**.983**

	Component			Communalities
Fe	-.147	**.942**	-.140	**.929**
% variance	58.4	20.6	7.6	**86.6%**

(a)

	Component				Communalities
	1	2	3	4	
pH	-.084	-.046	-.736	-.006	**.551**
Temperature	.451	-.249	.115	-.554	**.585**
EC	**.646**	-.005	.470	.204	**.680**
Colour	.081	**.822**	-.011	.071	**.687**
Turbidity	.060	**.831**	.328	-.049	**.803**
TS	**.622**	-.028	.195	**.695**	**.908**
TDS	**.844**	-.019	.342	.179	**.863**
TSS	.158	-.025	-.023	**.890**	**.818**
T-Hardness	**.934**	-.005	.161	.105	**.910**
T-Alkalinity	**.645**	.495	-.397	.045	**.821**
T-Acidity	.230	-.019	**.747**	-.133	**.629**
Cl	.249	-.519	.361	.229	**.514**
NO_3	-.065	-.385	-.492	**.586**	**.739**
Ca	**.952**	.071	.162	.111	**.949**
Mg	**.850**	-.047	.003	-.203	**.765**
Fe	-.471	.435	-.074	.048	**.419**
% variance	34.4	15.2	13.2	9.6	**72.8%**

(b)

T-hard-total hardness, T-Alkal-total alkalinity, T-acidity-total acidity,

Table 3. (a). The principal component analysis of groundwater samples at the urban site. (b). The principal component analysis of groundwater samples at the rural site.

This component source might be linked to soil source. Factor 3 (represents lowest portion of 7.6 % of the explained dataset) is moderately loaded for Ca and anti-correlated with temperature might suggest contamination by mineral dust e.g. cement from construction activities. Ca has always been linked with construction activities [33].

The rotated PCA at the rural site has identified four components that explained 72.8 % of the total variance. Factor 1 has significant loadings for EC, TS, TDS, total hardness, alkalinity, Ca and Mg. The factor resembles factor 1 in the PCA conducted for urban site. This factor might best describe organic pollutant source. Organic pollutants from the oil palm processing factory located in the rural site might have influenced the rural groundwater quality resulting in water contamination. In the factor 2 are positive significance of colour, turbidity and moderate loadings for total alkalinity and Fe. This component might suggest soil runoff. Factor 3 is highly associated with total acidity and moderate significance for EC. This factor might be related to the palm oil waste source. Factor 4 shows loadings for TS, TSS and NO_3. This might represent sources from land use. According to Hooda et al. [34], nitrate pollution in the rural groundwater is controlled by factors such as land use, climate and soil. Run-off from soil can also raise the solid levels of groundwater.

4. Conclusion

This study compared the groundwater quality of an urban site and a rural site in Nigeria. Results showed higher values of most of the observed parameters at the urban site relative to samples collected in the rural location. This showed that the groundwater qualities of rural wells are less contaminated compared to the urban groundwater. The effect of seasonal variation was also observed. The rotated PCA was conducted on groundwater samples from both study areas. The major contaminants in groundwater in urban areas were mixed pollution sources e.g. leachates from waste, effluents from industries, agricultural land and seepage from soak-away or septic tanks, soil and mineral dust. The varimax PCA revealed the major sources of pollution to groundwater in the study area. These include: organic, palm-oil wastes soil and land use. This study recommends further studies on heavy metals and polycyclic aromatic hydrocarbons (PAHs) in groundwater from rural and urban areas. Thorough treatment is required for groundwater samples from the urban areas before it could be certified fit for consumption.

Author details

A.M. Taiwo, A.T. Towolawi, A.A. Olanigan, O.O. Olujimi and T.A. Arowolo*

*Address all correspondence to: tarowolo@yahoo.com

Department of Environmental Management & Toxicology, Federal University of Agriculture, Abeokuta, Ogun State, Nigeria

References

[1] Lefort, R. (2006). Down to the last drop. UNESCO Sources.84, 7 pp.

[2] Orebiyi, E.O., Awomeso, J.A., Idowu, O.A., Martins, O., Oguntoke, O., and Taiwo, A.M. (2010). Assessment of pollution hazards of shallow well water in Abeokuta and Environs, Southwest, Nigeria. *American Journal of Environmental Sciences*. 6(1):50-56.

[3] Jain, C.K., Bhatia, K.K.S. and Vijay, T. (1995). Ground water quality monitoring and evaluation in and around Kakinada, Andhra Pradesh, Technical Report,CS (AR) 172, National Institute of Hydrology, Roorkee, 1994-1995.

[4] Alley, W. M. (1997). Regional groundwater quality. John Wiley & Sons. 364p.

[5] Kumar A. (2004). *Water Pollution*. Nisha Enterprises New Delhi Pp 33-35

[6] Gbadebo, A.M., J.A. Oyedepo, and A.M. Taiwo. (2010). Variability of nitrate in groundwater in some parts of southwestern Nigeria.*The Pacific Journal of Science and Technology* 11(2): 572-584.

[7] Ofodile, M.E. (2002). Groundwater study and development in Nigeria. University of Ibadan Press, Nigeria.

[8] Taiwo, A.M. (2012). Source identification and apportionment of pollution sources to groundwater quality in major cities in Southwest, Nigeria. *Geofizika* 29:157-174.

[9] McCasland, M., N.M. Trautmann, R.J. Robert and K.S. Porter (2007). Nitrate: Health effects in drinking water. http://psep.cce.cornell.edu/factsslides-self/facts/nit-heef-grw85.aspx

[10] WHO (2001). Cholera in Nigeria. http://www.who.int/csr/don/2001_11_27/en/. Accessed: 18/08/14.

[11] UNICEF. (2005). National Rural water supply and Sanitation investment programme. Final draft, pp: 1.

[12] Jones, A. A. and Hockey, R. A. (1964).The geology of part of southern Nigeria. *Geological Bulletin* 31, 101 pp.

[13] Oyedele, K. F., Ogagarue D. O and Esse, O. (2011). Groundwater potential evaluation using surface geophysics at Oru-Imope, South-Western Nigeria. *European Journal of Scientific Research*63: 515–522.

[14] Adelana, S. M. A., Olasehinde, P. I., Bale, R. B., Vrbka, P., Edet, A. E. and Goni, I. B. (2008). An overview of the geology and hydrogeology of Nigeria. In *Applied Groundwater Studies in Africa, IAH Selected Papers on Hydrogeology, edited by Adelana S. and MacDonald, A.* CRC Press/Balkema, Taylor & Francis Group, The Netherlands, 13, 171–199, Print ISBN: 978-0-415-45273-1, eBookISBN: 978-0-203-88949-7, doi: 10.1201/9780203889497.ch11.

[15] Vogel, A.I., 1976. A Textbook of inorganic analysis, 3rd Edition, Longman, London.

[16] Ademoroti, C. A. (1996). *Standard methods for water and effluents analyses*.Environmental Chemistry and Toxicology. Ibadan, Nigeria: Foludex Press Ltd.

[17] EPA (2003). Environmental chemistry, transport and fate. http://www.epa.gov/raf/metalsframework/pdfs/chaper3.pdf. Accessed 18/08/14.

[18] Taiwo, A. M., Gbadebo, A. M. And Awomeso, J.A. (2010a). Potability assessment of selected brands of bottled water in Abeokuta, Nigeria. *Journal of Applied Science and Environmental Management* 14 (3): 47-52. www.ajol.info/index.php /jasem/article/view/ 61464/49604.

[19] Jarup, L. (2003). Hazards of heavy metal contamination. *British Medical Bulletin* 68(1): 167-182.

[20] World Health Organization (2008). Guidelines for drinking water quality, Geneva WHO 2008.

[21] http://vortex.weather.brockport.edu/~jzollweg/oakorchard/docs/waterquality.pdf.

[22] Awomeso, J. A., Taiwo, A. M., Morawo, O. A. and Moyosore., J. O. (2011). Possible Abstraction Sites along Oshun River-Lower Course in Ogun and Lagos State, Nigeria for Sustainable Production of Quality Potable Water. *Journal of Science and Technology* 31 (3): 58-67.

[23] FAO (2012).Water quality monitoring, standards and treatment. In: *Fishery Harbour Manual on the Prevention of Pollution-Bay of Bengal Programme.* FAO repository document. http://www.fao.org/docrep/x5624e/x5624e05.htm. Accessed 18/08/14.

[24] http://www.env. gov.bc.ca/wat/wq/BCguidelines/colour/colour_over.html

[25] Mento d I'exploitant de leauet del I'assainissesment. (1980). Lyonnaise Deseaux, p. 273-283.

[26] Awomeso, J. A., Taiwo, A. M. and Adenowo, J. A. (2010). Studies on the Pollution of Water Body by Textile Industry Effluents in Lagos, Nigeria. *Journal of Applied Sciences in Environmental Sanitation* 5 (4): 337-343.

[27] Roberts, R. J. (1978). Fish Pathology. Bailliere-Tindall, London, UK.

[28] Taiwo, A.M., A.O. Adeogun, K.A. Olatunde, and K.I. Adegbite. (2010b). Analysis of groundwater quality of hand-dug wells in peri-urban area of Obantoko, Abeokuta, Nigeria for selected physico-chemical parameters.*Pacific Journal of Science and Technology* 12(1):527-534.

[29] Satyanarayana, P., Raju, N.A., Harikrishna, K and Viswanath, K. (2013). Urban Groundwater Quality Assessment: A Case Study Of Greater Visakhapatnam Municipal Corporation Area (Gvmc), Andhra Pradesh, India. *International Journal of Engineering Science Invention* 2 (5): 20-31.

[30] Ayedun, H., Taiwo, A. M., Umar, B. F., Oseni, O. A. andOderinde, A. A. (2011). Potential Groundwater Contamination by Toxic metals in Ifo, Southwest Nigeria. *Indian Journal of Science and Technology* 4 (7); 820-823.

[31] Adekunle, I.M., Adetunji, M.T., Gbadebo, A.M. and Banjoko, O.B. (2007). Assessment of groundwater quality in a typical rural settlement in southwestNigeria. *International Journal of Environment and Public Health 4 (4)*: 307-318

[32] Ullah, R., Malik, R.N. and Qadir, A. (2009). Assessment of groundwater contamination in anindustrial city, Sialkot, Pakistan. *African Journal of Environmental Science and Technology* 3 (12): 429-446. http://www.academicjournals.org/AJEST

[33] Watson, J. G., Chow, C. (2007). Receptor models for particle source apportionment of suspended particles. *In: Introduction to Environmental Forensics,* 2nd*edn. Murphy, B. and Morrison, R. (Eds).* Academic Press: New York, p. 273-310.

[34] Hooda, P.S., Edwards, A.C, Anderson, H. A., Miller, A. (2000). A review of water quality concerns in livestock farming areas. *Scienc*e of the Total Environment 250: 143.

Chapter 8

Variability in Heavy Metal Levels in River Water Receiving Effluents in Cape Town, South Africa

O.O. Olujimi, O.S. Fatoki, J.P. Odendaal and
O.U. Oputu

Additional information is available at the end of the chapter

http://dx.doi.org/10.5772/59077

1. Introduction

One of the most critical problems of developing and developed countries is improper management of vast amount of wastes generated by various anthropogenic activities. Though, very pronounced in the developing countries due to availability of potable water sources. More challenging is the unsafe disposal of these wastes into the ambient environment. Water bodies especially freshwater reservoirs are the most affected. This has often rendered these natural resources unsuitable for both primary and/or secondary usage [1]. Water shortage is an important concern in arid areas such as Africa, Southern Asia and Middle East and even in some parts of the World which it may lead to a war crisis [2].

On the other hands continued population growth, increased per capital water consumption and increased water requirements for industry and irrigation result in considerable decrease of usable water resources [3]. Therefore, treated wastewater recycling into the hydrological cycle is of significant importance and has many benefits. The major uses of treated wastewater are in agricultural irrigation, industrial activities and groundwater recharge. With respect to public health, principles of engineering economy, aesthetic standards and more importantly public acceptance, wastewater reuse can be developed.

However, incomplete removal of organic compounds and heavy metals from treated effluents can cause long term effects on the ecosystem even when the impact is not immediately feasible [4-6]. Although, a number of studies have been conducted on heavy metals in river in association with intensive farming and industrial activities in South Africa, most especially in the Guateng Province, no study has reported levels of heavy metals in relation to wastewater treatment plants in Cape Town. Thus, the main objectives of this study were to assess: (i) levels of heavy metals in river water receiving treated effluents from wastewater treatment plants

(ii) identification of the possible point source pollution of heavy metals from wastewater treatment plant if any and (iii) compare if reported levels are in compliance with the South Africa and other guidelines for freshwater management.

2. Materials and method

2.1. Methods

All the determinations were carried out by Inductively Coupled Plasma Mass Spectrometry (ICP-MS) located at the geology Department, University of Stellenbosch. The Agilent 7700 instrument was used with a Meinhardt nebulizer and silica cyclonic spray chamber with continuous nebulization. The operation parameters are: Plasma RF power: 1550 W; Sample depth: 8.0 mm; Carrier gas: 1.08 L/min; Nebulizer pump: 0.10 rps; Helium gas: 5.3 $mlmin^{-1}$ for ICPMS. The isotopes of the elements determined were: ^{52}Cr, ^{59}Co, ^{60}Ni, ^{63}Cu, ^{111}Cd, ^{75}As, ^{208}Pb, ^{202}Hg, ^{66}Zn.

2.2. Reagents

Water (resistivity 18.2 MΩ cm) was de-ionized by use of a Milli-Q system (Millipore, Bedford, MA, USA). Certified standard of all the metals (As, Cd, Cr, Co, Cu, Pb, Ni Hg and Zn) to cheek for instrument performances and $AuCl_3$ were obtained from Merck, South Germany. Ultrapure nitric acid (65 %) and 32 % hydrogen peroxide were obtained from Fluka Kamika, Switzerland. 1000 mgL^{-1} of metal stock standard solution (As, Cd, Cr, Co, Cu, Pb, Ni Hg and Zn) was supplied by Sigma-Aldrich.

2.3. Study areas

Final effluent (at the discharge point) of six wastewater treatment plants namely; Athlone, Bellville (which consist of the Old and New plants), Kraaifontein, Potsdam, Stellenbosch and Zandvliet) were investigated for heavy metals. Five of these WWTPs were located in the City of Cape Town, while one is located in Stellenbosch. Rivers associated with each treatment plant are: Athlone-Vygekraal River; Bellville-Kuils River; Kraaifontein-Mosselbank River; Potsdam-Diep River; Zandvliet-Kuils River and Stellenbosch-Veldwachters River. All the sampled WWTPs receive wastewater from both domestic and industrial effluents, except kraaifontein that receives mainly (about 90 %) domestic wastewater. Samples were taken at the point of discharge, as well as upstream and downstream from point of discharge (about 1-2km) to evaluate the possible impact of effluent on heavy metals and organic compounds load on the aquatic.

2.4. River water collection and digestion

Samples were collected from eighteen sampling sites consisting of upstream, discharge point, downstream and a control site (Kirstenbosch Botanical Garden). Samples were collected in 1litre plastic container which were initially washed with detergent and rinsed with distilled water. The containers were finally soaked in 10 % Nitric acid. The containers were then rinsed at least three times with MilliQ water. At the sampling sites, containers were rinsed three times with the water samples before being filled with the samples. The samples were preserved by adding few drops of conc. HNO_3 to each sample bottle and the pH adjusted to 2.0 by the use of pH meter. The samples were transferred on ice chest to the laboratory prior to storage in a refrigerator at about 4°C before analysis. As samples may contain particulate or organic materials, pretreatment in the form of digestion is required before analysis. Nitric acid digestion was employed [7]. A few drops of $AuCl_3$ were added to 100 mL of unfiltered river water samples to keep Hg ion in solution prior to digestion. Water sampling for heavy metals analysis commenced in January 2010 and ended in December 2010.

2.5. Quality control

The analytical data quality was guaranteed through the implementation of laboratory quality assurance and quality control methods, including the use of standard operating procedures, calibration with standards, analysis of reagent blanks, recovery of known additions and analysis of replicates. All the analyses were carried out in triplicate and the results were expressed as the mean. The instrument calibration was checked with SRM 1643a (Trace elements in water) purchased from NIST, Gaithersburg, USA. The instrument reproducibility was check using in-house prepared drift standard (1 $\mu g\ L^{-1}$ of all the trace and rare earth elements and 1 $mg\ L^{-1}$ of Na, K, P, Ca and Mg). The elemental concentrations and accuracy of the certified reference materials SRM 1643a. The instrument drift was very negligible as measurement gave a ratio of 0.89 to 1.05. The result of the SRM 1640a (Trace elements in water) was acceptable to validate the calibration.

3. Result and discussion

3.1. Arsenic

In this study, the seasonal concentrations of arsenic in water of the selected river systems receiving wastewater effluent were determined for samples taken from points about 1-2 km up and downstream from the point of final discharge. The range of the annual mean of arsenic in water for all sampling sites in comparison with other studies is presented in Table 1. The graphical forms of the seasonal variation at each sampling point for water is presented in Figure 1. The average levels of arsenic in water samples obtained from the river system ranged from 0.56 μgL^{-1} to 23.78 μgL^{-1} for the nineteen sampling points. The highest level of arsenic was obtained at sampling point 7 (Bellville WWTP downstream) during winter and the lowest at sampling point 12 (Stellenbosch WWTP discharge point) as depicted in Figure 1. The annual mean concentration of arsenic from each sampling point ranged from 1.62 μgL^{-1} (Site 1) to 13.7 μgL^{-1} (Site 13). The seasonal trend of arsenic in water shows that the summer samples had the least concentration while the winter had the highest concentration for most of the sampling sites except for sites 8, 11 and 15. Studies in several countries reported levels of arsenic in water ranging from 1.25 μgL^{-1} to 5114 μgL^{-1} [8-17] (Table 1). When comparing the findings of this study with other reported values, it was obvious that the result of this study was generally low except for sites 7, 11 and 13 where reported values were higher than the South Africa water quality guidelines. Reported concentrations were within the human consumption (except for 7, 11 and 13), livestock watering, irrigation and aquaculture uses [18,19]. Generally, the wastewater treatment plants are believed to be one of the possible routes of organic and inorganic pollutants into the river systems. However, from this study, the annual mean values for arsenic at the discharge point was lower compared to the upstreams and downstreams values of the river, but higher than the values at the control site (Site 1). The high concentrations of arsenic at site 7 may be attributed to defeacating by cattle in the water as the water is used for livestock management in the area. Another possible means of arsenic in this section of the river may be attributed to the use of sodium salt of arsenous acid to treat tick infestations on cattle [20] and waste tyres dump. At sites 11 and 13, the high concentration of arsenic recorded may be attributed to seepage of landfill leachate into the river systems at site 11. The high concentration at site 17 may be attributed to channelization of the upstream and informal settlement around the sampling point. There is also possibility of storm water contamination as many rivers in Cape Town are known to receive storm water carrying industrial effluents, wastes from home and farms or seepage from groundwater [21]. Sites 7 (Bellville wastewater downstream) and 14 (Zandvliet wastewater upstream) are sampling points on Kuils River. Site 7 is located far upstream of site 14 which is about 2 km of Zandvliet point of discharge. High arsenic level at this portion of this river may be due to storm and wastewater effluent from the biggest informal settlement in Cape Town (Khayelitsha) with over 1.2 million inhabitants.

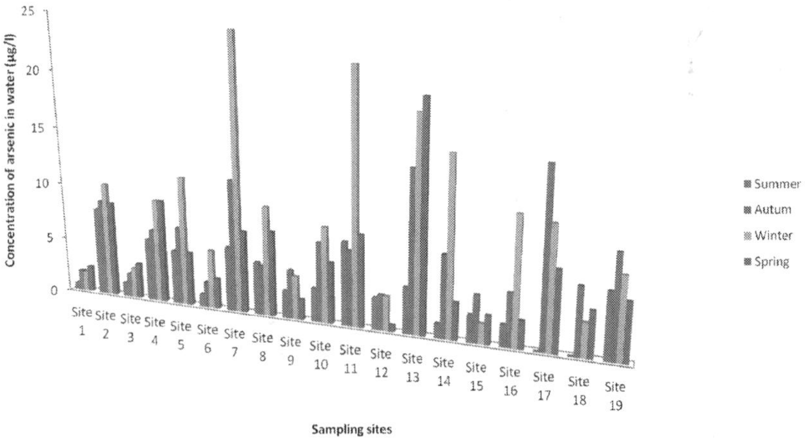

Figure 1. Seasonal trend in arsenic concentration (µgL^{-1}) in river water receiving waste effluent from WWTPs Site 1: Kirstenbosch Botanical garden (Control Site); Site 2: Potsdam WWTP upstream; Site 3: Potsdam WWTP discharge point; Site 4: Potsdam WWTP downstream; Site 5: Bellville WWTP upstream; Site 6: Bellville WWTP discharge point; Site 7: Bellville WWTP downstream; Site 8: Kraaifontein WWTP upstream; Site 9: Kraaifontein WWTP discharge point; Site 10: Kraaifontein WWTP downstream; Site 11: Stellenbosch WWTP upstream; Site 12: Stellenbosch WWTP discharge point; Site 13: Stellenbosch WWTP downstream; Site 14: Zandvliet WWTP upstream; Site 15: Zandvliet WWTP discharge point; Site 16: Zandvliet WWTP downstream; Site 17: Athlone WWTP upstream; Site 18: Athlone WWTP discharge point; Site 19: Athlone WWTP downstream.

http://dx.doi.org/10.5772/59077

		Range	Mean (Mean ± SD)	Reference
Water	WHO		10	
	Cape Town	**1.62-13.70**	-	**this study**
	Pakistan	-	97.5±28.5	[16]
	Tuskegee Lake	-	0.06±0.23	[12]
	China	<100-1860	-	[14]
	Si Thammarat Province, Thailand	1.25-5114	503.4	[9]
	Taiwan	up to 1800	-	[8]
	Bangladesh	<2->900	-	[10]
	Taiwan	1-50	21±10	[11]
	Japan	0.24-0.68	0.48	[13]
	China	8.33-28.81	-	[17]
	Mexico	70-160	-	[15]

Table 1. Concentration of arsenic in river water (μgL^{-1}) in comparison with other globally published values

3.2. Cadmium

Seasonal concentrations change of cadmium in water of the river systems receiving wastewater effluents and Kirstenbosch Botanical Garden are presented in graphical form (Figure 2). For all the sites investigated, the average mean concentrations of Cd in water samples obtained from the river systems ranged from 0.09 μgL^{-1} to 14.78 μgL^{-1} for the 19 sampling points as placed in Figure 2. The highest level of cadmium in water was obtained at Site 17 (Athlone WWTP Upstream) during the autum sampling season and the lowest at Site 14 (Zandvliet WWTP Upstream) during autum. The annual average cadmium concentration found in this study ranged from 1.44 μgL^{-1} Site 15 (Zandvliet WWTP discharge point) to 7.96 μgL^{-1} Site (17 Athlone WWTP downstream). In previous study conducted in South Africa, Fatoki *et al.* [21] reported concentration range of 0.01 to 26 mgL^{-1}, while another study [22] reported concentration range of between 2 and 4 μgL^{-1}. Cadmium concentration had not been previously reported in the selected river systems in Cape Town as attention had been focus on other toxic metals and especially in sediment and soil samples. Similarly, in another study [23], cadmium was detected at about 6 μgL^{-1} for upstream and downstream samples collected in the Eerste River for two sampling seasons. Elsewhere in South Africa, it was reported that levels of cadmium in water ranged from 1.6 μgL^{-1} to 260 μgL^{-1} as placed in Table 2 [21-28]. Annual values reported in this study were lower compared to previous finding in the Eastern Cape and Nigeria (Table 2). Cd concentrations in non-polluted natural waters usually are lower than 1 μgL^{-1}, have been reported. On comparison with South Africa water quality guidelines, the reported levels of cadmium indicated that all sampling sites concentration were within the limits for human consumption except for site 17 and 19 while all sites, 17 and 19 inclusive were below the set limits of 10 μgL^{-1} for livestock watering and irrigation of farmlands. However, in relation to protection of aquatic life's, reported concentrations for all the 19 sites were above the 0.2 μgL^{-1} and 0.017 μgL^{-1} limits by DWAF [18] and CCME [19] respectively.

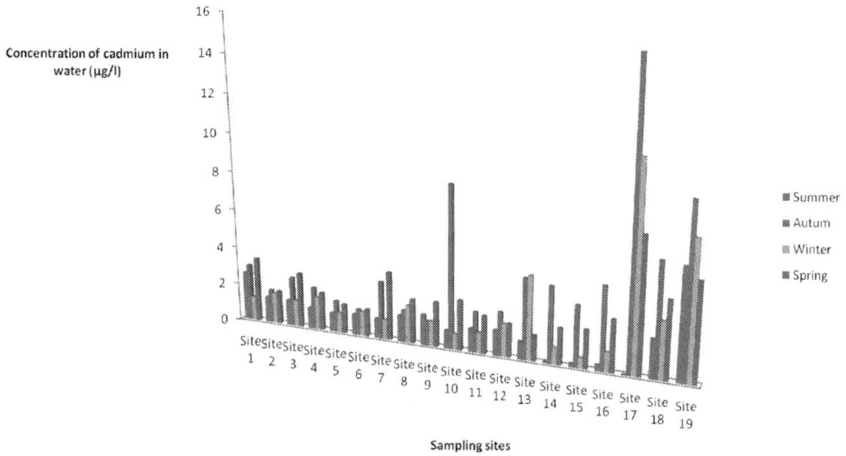

Figure 2. Seasonal trend in Cd concentrations (µgL^{-1}) in river water receiving waste effluent from WWTPs Sites are the same as listed in Figure 1

		Range	Mean (Mean ± SD)	Reference
Water	South Africa	2-4	-	[22]
	South Africa	10-260	-	[21]
	Cape Town	**1.44-7.96**	-	**this study**
	South Africa	30-44	-	[24]
	Mexico	-	14	[15]
	Egypt	<0.5-34	25	[28]
	India	0.67-5.61	-	[27]
	South Africa	-	6	[23]
	South Africa	1.6-9.3	-	[25]
	Nigeria	30-390	-	[26]

Table 2. Concentration of Cd in river water (µgL^{-1}) and comparison with other globally published values

3.3. Chromium

Results of seasonal concentration of chromium in both the river water and sediment are presented in graphical forms as depicted in Figure 3. The average chromium concentrations ranged from 9.27 µgL^{-1} to 327.29 µgL^{-1}. The highest concentration was at site 16 during summer while the least was at site 12 (Potsdam WWTP upstream) during the spring. The annual mean concentration in water ranged from 16.19 µgL^{-1} (Potsdam WWTP upstream) to 206.57 µgL^{-1} (Site 8, Kraaifontein Upstream). To the best of our knowledge, no work had reported Cr levels in selected river systems in Cape Town. Aside from Nigeria and Mexico, reported annual concentration ranges were higher than values reported in Egypt, Greece and China [15,26,29-31] (Table 3). The presence of Cr (III) in drinking water is unlikely due to low solubility of the hydrated Cr (III) oxide. The more stable Cr (VI) may occur especially in the vicinity of industries which result in environmental pollution. The Target Water Quality Range (TWQR) for aquatic ecosystem is 7 μgL^{-1} while the human consumption target is 50 μgL^{-1} [18]. The average annual concentration of chromium for the sites exceeded the TWQR guideline for aquatic ecosystem while sites 1, 2 and 5 were within the 50 μgL^{-1} limits for human consumption. The high concentration of Cr in the river systems may be due to high number of vehicle repair workshops, electro plating industries and paint industries in the City of Cape Town, as their waste effluents may enter the rivers as storm water. Also, from this study, a major route of Cr to the river systems in Cape Town and Stellenbosch are through wastewater treatment plants effluents and landfill site leachate (Figure 3). For all sites, Cr values also exceeded the recommended value of 2 μgL^{-1} for aquacultural uses, while all sites except for sites 6, 8, 10, 11, 13, 16 and 19 are within the TWQR for irrigation purposes (100 μgL^{-1}) but within the livestock watering guidelines. However, comparing with international standards, the reported values in this study exceeded the 8 μgL^{-1} and 50 μgL^{-1} for irrigation water and livestock water use [19].

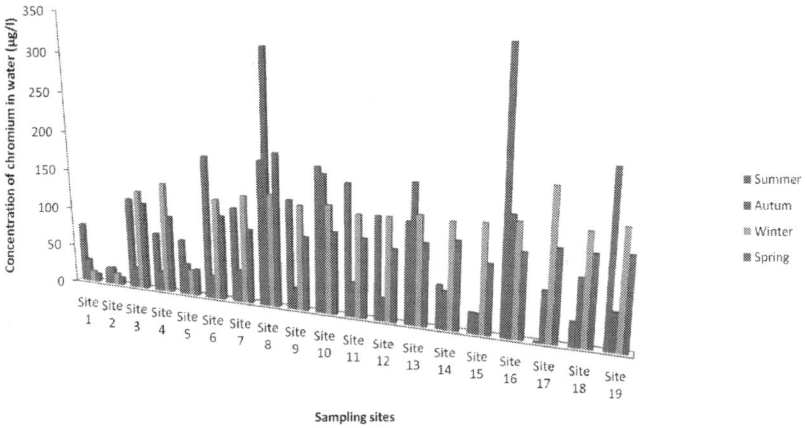

Figure 3. Seasonal trend in Cr concentrations (μgL^{-1}) in river water receiving waste effluent from WWTPs; Sites are the same as listed in Figure 1

		Range	Mean (Mean ± SD)	Reference
Water	**Cape Town**	**16.19-206.57**	-	**this study**
	Nigeria	0.00-270	-	[26]
	Mexico	60-212	-	[15]
	China	6.67-20.44	-	[30]
	Greece	1.02-4.58	-	[29]
	Egypt	8-88	-	[31]

Table 3. Annual concentration (Mean) of Cr in river water (μgL^{-1}) and comparison with other globally published values

3.4. Cobalt

The seasonal variation in Co concentrations from all the 19 sampling sites is presented in Figure 4. The graphical presentation shows that Co ranged from 0.15 μgL^{-1} to 4.95 μgL^{-1}. The highest concentration of Co was obtained at sampling site 2 (Potsdam WWTP upstream) during spring and the lowest was obtained at site 16 (Zandvliet WWTP downstream) during winter. The annual mean of Co concentration at each sampling site ranged from 0.96 μgL^{-1}(Site 15, Zandvliet WWTP discharge point) to 3.66 μgL^{-1} (Site 2, Potsdam WWTP upstream) (Figure 4). The values reported in this study were considerably lower when campared to previous studies in South Africa and elsewhere [15,24,28,32] (Table 4). Cobalt is considered an essential metal and form part of Vitamin B12, which is useful during the synthesis of red-blood cell. Ingestion of cobalt at concentration higher than 2000 μgL^{-1} may result in chronic human effect [24]. The reported concentration of cobalt in this study exceeded the unpolluted surface water quality guidelines [18]. However, the water is suitable for agricultural and livestock watering purposes.

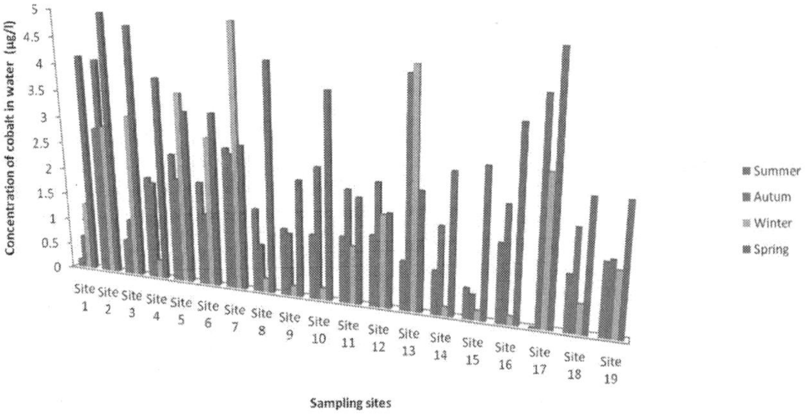

Figure 4. Seasonal trend in Co concentrations (μgL^{-1}) in river water receiving waste effluent from WWTPs; Sites are the same as listed in Figure 1

		Range	Mean (Mean ± SD)	Reference
Water	South Africa	0.04-1.99	-	[32]
	Cape Town	**0.96-3.66**	-	**this study**
	Egypt	<5-43	-	[28]
	Mexico	2-8	-	[15]
	China	0.42-6.18	-	[30]
	South Africa	trace-62	-	[24]

Table 4. Concentration of Co in river water (μgL^{-1}) and comparison with other globally published values

3.5. Copper

The average concentrations of copper in water samples of the selected river system are graphical form as shown in Figure 5. The average levels of Cu in water samples obtained from the 19 sampling points ranged from 6.99 μgL^{-1} to 305.39 μgL^{-1}. The highest level of copper was obtained at sampling site 11 (Stellenbosch upstream) during autum and the lowest at sampling point 1 (control site, Kirstenbosch botanical garden) during summer as depicted in Figure 5. The annual mean of copper concentration at each sampling site ranged from 18.23 μgL^{-1} (Site 9, Kraaifontein discharge point) to 120. 52 μgL^{-1} (Site 14, Zandvliet upstream). Previous study on Eerste River [23] reported concentration range of 60-70 μgL^{-1} while studies elsewhere in South Africa reported Cu concentration of 2-530 μgL^{-1} [22,24,25] (Table 5). Copper concentration at Site 11 during autum season may be attributed to leachate seepage into the river system and the dumping of the demolition material coupled with storm water from the landfill site. Levels at site 14 may be attributed to the closeness to an informal settlement. Reported Cu concentration were lower compared to studies elsewhere (Table 5). The annual average values in this study were within the South African water quality guideline for Cu in domestic water usage (DWAF, 1996). The TWQR limits for irrigation and livestock watering are 200 μgL^{-1} and 5000 μgL^{-1} with chronic impact on livestock expected between 1000 and 10,000 μgL^{-1} depending on the livestock [18]. Cu concentrations reported in this study were within these limits except for Site 11 (Stellenbosch upstream) during the autum season. Generally, all the sampling sites values for Cu exceeded the set limits of 0.3 μgL^{-1} for the protection of aquatic life. Wastewater treatment plants shows to be one of the major routes of copper into the freshwater systems from this study.

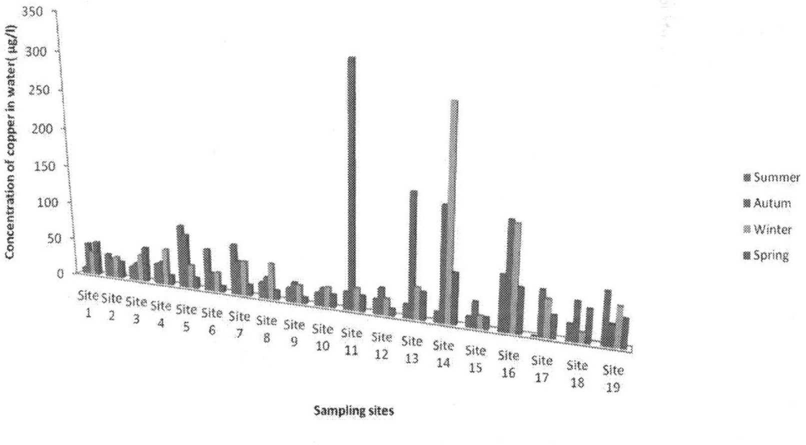

Figure 5. Seasonal trend in Cu concentrations (μgL^{-1}) in river water receiving waste effluent from WWTPs; Sites are the same as listed in Figure 1

		Range	Mean (Mean ± SD)	Reference
Water	**Cape Town**	**18.23-120.52**	-	**this study**
	South Africa	383-387	-	[24]
	Mexico	>200	-	[15]
	India	102.82-258.7	161.43	[27]
	South Africa	60-70	-	[23]
	South Africa	2-3	-	[25]
	Egypt	22-54	-	[31]
	Nigeria	0.00-1536	-	[26]
	South Africa	100-530		[21]

Table 5. Concentration of Cu in river water (ugL^{-1}) and comparison with other globally published values

3.6. Lead

The result of seasonal concentrations of lead in water and sediment of the selected river systems receiving wastewater effluent are presented in Figure 6. The average values of Pb in water samples obtained from the river system ranged from 4.18 μgL^{-1} to 86.73 μgL^{-1} for the 19 sampling points as shown in Figure 6. The highest level of lead was obtained at Site 16 (Zanvliet WWTP point of discharge) during summer and the lowest at Site 1(control site, Kirstenbosch Botanical Garden) during summer. Meanwhile, the annual mean value of lead at each sampling site in this study for water ranged from 17.6 μgL^{-1} to 52.9 μgL^{-1}. Previous studies in South Africa had reported Pb concentration ranging below detection limit to 1110 μgL^{-1} [21,23-26,28,29] (Table 6). Meanwhile, another study Reinecke *et al.* [23] reported 30 to 40 μgL^{-1} of lead in the Eerste River. Effluent discharges from sewage treatment plant and industries had been suggested as possible routes of Pb into river systems. Thus, considering the values reported in the study, wastewater effluent is a factor to high lead concentration in the river system. Though, the study shows that the final effluent concentration were generally low for lead, and the effluent helps to further dilute the river water concentration, possible contamination source could not be ruled out. The recommended threshold level of lead for South Africa Rivers is 10 μgL^{-1} [18]. The results shows that the annual average value of lead for all the sampling points of the river system and the control site were above the TWQR threshold level for human consumption and aquacultural purposes. However, reported values were within the TWQR for irrigation and livestock watering. The water is unsuitable for the protection of aquatic ecosystems as TWQR limits of 0.2 μgL^{-1} was exceeded.

3.7. Mercury

In this study, the seasonal concentrations of mercury in water of the selected river system and control site are depicted in Figure 7. The average levels of Hg in water samples obtained from the 19 sampling sites ranged from 0.1 μgL^{-1} to 8.09 μgL^{-1} while the annual mean concentration for each sampling site ranged from 1.45 μgL^{-1} to 2.58 μgL^{-1} The highest level of mercury was obtained at sampling site 15 (Zandvliet discharge point) during the spring season and the lowest at sampling point 2 (Potsdam WWTP upstream) as depicted in Figure 7. Previous study in Eastern Cape had reported concentration of Hg 0.003 mgL^{-1} [33]. While Retief *et al.* [32]

reported Hg concentration range of 0.125 μgL^{-1} to 0.513 μgL^{-1} in the Vaal dam, South Africa. Previous studies in several countries reported levels of mercury in water were ranged from not detected to 1502 μgL^{-1} [27,34-37] (Table 7). The recommended TWQR threshold level of mercury for South African rivers for human consumption is 1.00 μgL^{-1} [18]. The average values of mercury for all the samplings sites exceeded the limits, though there are instances during sampling period where Hg concentrations were below this guideline. Also, Hg concentration exceeded TWQR guideline for the protection of aquatic ecosystem, livestock watering and aquaculture uses. Considering the effect of ingesting Hg through the river water, the water system is unsafe for domestic, agricultural, livestock and aquaculture uses.

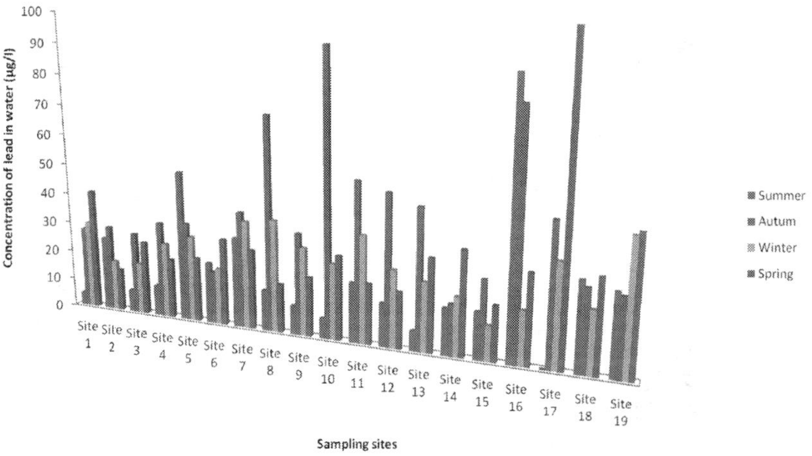

Figure 6. Seasonal trend in Pb concentrations (μgL^{-1}) in river water receiving waste effluent from WWTPs; Sites are the same as listed in Figure 1

		Range	Mean (Mean ± SD)	Reference
Water	**Cape Town**	**17.64-52.99**	-	**this study**
	Nigeria	10-2570	-	[26]
	South Africa	10.5-20.1	-	[25]
	Greece	ND-12.61	-	[29]
	South Africa	30-40	-	[23]
	South Africa	24-350	-	[24]
	South Africa	240-1110	-	[21]
	Egypt	5-57	-	[28]

Table 6. Concentration of Pb in river water (μgL^{-1}) and comparison with other globally published values

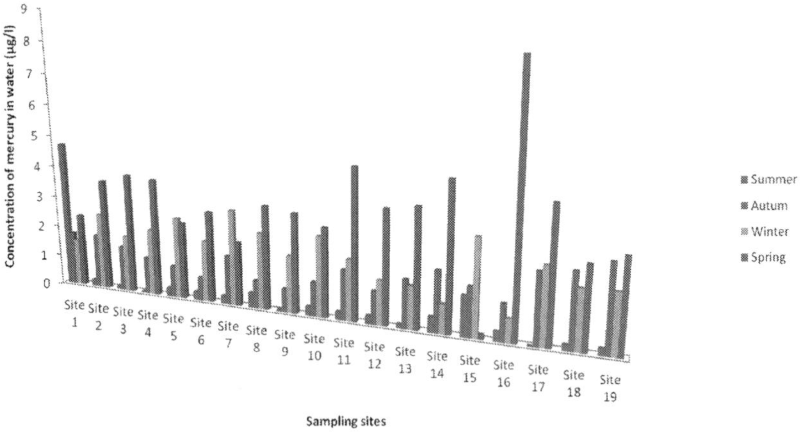

Figure 7. Seasonal trend in Hg concentration (µgL^{-1}) in river water receiving waste effluent from WWTPs; Sites are the same as listed in Figure 1

		Range	Mean (Mean ± SD)	Reference
Water	**Cape Town**	**1.45-2.58**	-	**this study**
	India	0.015-0.194	0.001	[27]
	South Africa	trace-3	-	[33]
	Spain	ND	-	[37]
	Turkey	1.50-1502	-	[36]
	Spain	ND-2.09	-	[35]
	Spain	2.80-5.70	-	[34]

Table 7. Concentration of Hg in river water (µgL^{-1}) and comparison with other globally published values

3.8. Nickel

Seasonal concentrations of nickel in water, from all the 19 sampling locations are presented in Figure 8. The seasonal concentration ranged from 7.7 µgL^{-1} to 159.17 µgL^{-1}. The highest level of nickel was obtained at sampling point 3 (Potsdam discharge point) during winter and the lowest at site 15 (Zandvliet discharge point) during winter. Meanwhile, the annual average nickel concentration found in this study in the water samples ranged from 27.62 µgL^{-1} to 106.39 µgL^{-1} for Site 1 (Kirstenbosch Botanical Garden) and Site 3 (Potsdam discharge point), respectively. A study by Awofolu *et al.* [24], reported concentration of nickel found in Eastern Cape river to ranged from 201 µgL^{-1}to 1777 µgL^{-1}. Besides that, Retief *et al.* [32] reported a nickel concentration range of 2.89 µgL^{-1} to 27.2 µgL^{-1} in Vaal dam, South Africa (Table 8). Studies in several countries reported levels of nickel in water ranging from < 5 µgL^{-1} to 300 µgL^{-1} [15,24,

28,31,32] There was no water quality guidelines set by South Africa Department of Water Affair and Forestry for human consumption, protection of aquatic ecosystem and for aquacultural uses. However, the reported concentrations in this study were still within the TWQR of 200 μgL^{-1} and 1000 μgL^{-1} for irrigation and livestock watering. From this study, WWTP acts as one of the major routes of nickel into the freshwater system as concentration downstream of the treatment plants was higher than concentration upstream. This also established the anthropogenic route of nickel introduction into the environment

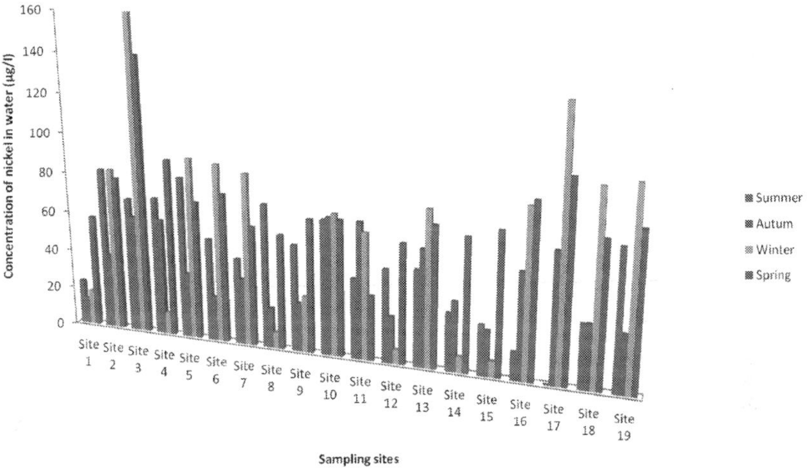

Figure 8. Seasonal trend in Ni concentration (μgL^{-1}) in river water receiving waste effluent from WWTPs; Sites are the same as listed in Figure 1

		Range	Mean (Mean±SD)	Reference
Water	**Cape Town**	**27.62-106.39**	-	**this study**
	Mexico	160-300	-	[15]
	South Africa	201-1777	-	[24]
	China	0.14-6.13	-	[30]
	Egypt	<5-33	-	[28]
	South Africa	2.89-27.16	-	[32]

Table 8. Concentration of Ni in river water (μgL^{-1}) and comparison with other globally published values

3.9. Zinc

Seasonal variation in the concentration of Zn in water samples from all the 19 sampling sites is presented in Figure 9. The average seasonal concentration ranged from 25.15 μgL^{-1} to 909.38 μgL^{-1}. The highest level of zinc was obtained at sampling site 15(Zandvliet discharge point) during summer and the lowest at sampling site 12 (Stellenbosch discharge point). Meanwhile, the annual mean zinc concentration found in this ranged from 172.79 μgL^{-1} (Site 1, Kirstenbosch Botanical Garden) to 722.07 μgL^{-1} (Site 13, Stellenbosch downstream). Previous study in the Western Cape Province had reported various concentration of Zn in river water. Jackson *et al.* [38], reported zinc concentration ranging from 100 μgL^{-1} to 2100 μgL^{-1} in Berg River and Jackson *et al.* [39] reported concentration range of between 100 μgL^{-1} and 4400 μgL^{-1}for studies conducted on Plankenburg and Diep Rivers. However, studies elsewhere in South Africa had reported concentration range of 10 μgL^{-1} to 43 μgL^{-1} [21,23,24,33] (Table 9). Meanwhile, studies in several countries reported levels of zinc in water were ranged from <5 μgL^{-1} to 97 μgL^{-1} [22, 23,26,27,39-46] (Table 9). The reported values in this study were lower compare to previous studies in Cape Town. Aside from the geology of the catchment, zinc concentration in the river systems pointed towards WWTPs and storm water carrying both industrial and domestic effluents. The recommended TWQR for Zn in water for domestic purposes is 3000 μgL^{-1} [18]. Thus, from the reported values, no health effect is expected from domestic use of the water from the sampling sites. However, the TWQR for the protection of aquatic ecosystem, aquaculture purposes, livestock watering and irrigation of are 2 μgL^{-1}, 30 μgL^{-1}, 0 to 20 mgL^{-1} and 100 μgL^{-1}. From this study, water from the river systems and the control site is not suitable for the protection of aquatic ecosystem or use for aquaculture purposes.

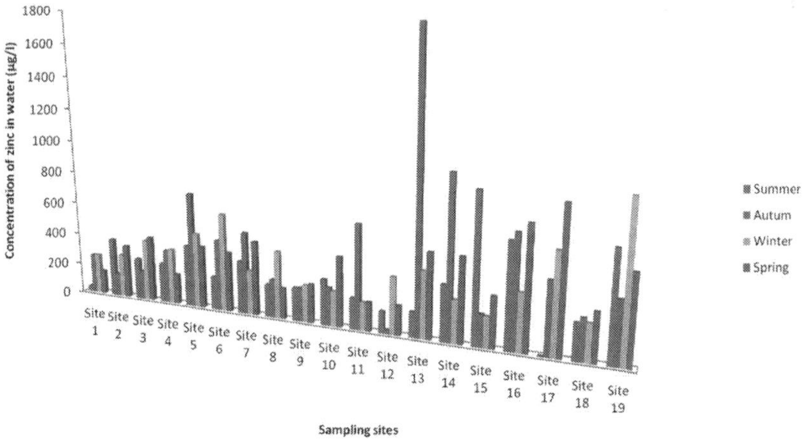

Figure 9. Seasonal trend in Zn concentrations (μgL^{-1}) in river water receiving waste effluent from WWTPs; Sites are the same as listed in Figure 1

Water		Range	Mean (Mean±SD)	Reference
	China	49.00-1142	-	[43]
	Cape Town	**144.46-1443.13**	-	**this study**
	China	29.6-121.68	86.82±30.05	[46]
	Denmark	362-1240	-	[42]
	Egypt	146.6-742	-	[28]
	South Africa	578-15240	-	[22]
	Turkey	39.02-75.31	-	[41]
	Uganda	44.27-1968.43	-	[45]
	India	128.66-308.02	210.10	[27]
	India	8.47-343.47	-	[40]
	South Africa	269.5-1081.2	-	[39]
	South Africa	44.7-77.5	-	[23]
	South Africa	2.9-211.3	-	[44]

Table 9. Concentration of Zn in river water (μgL^{-1}) and comparison with other globally published values

4. Conclusion

In the river water, arsenic and cadmium were within the normal level for human consumption but exceeded the limits for the aquatic life protection. Also, lead and mercury exceeded both limits for human consumption and sustainable aquatic life's. The trend in the levels of metals and arsenic in the river systems showed that the upstream and downstream are more polluted compared to the WWTP discharge points. This is an indication that the WWTPs might not completely be the pollution source of the river systems in the City of Cape Town. The reported trend may be attributed to waste dumping on the river course, indiscriminate wastewater discharge from industries, storm water runoff from agricultural lands and grey and domestic wastewater.

Author details

O.O. Olujimi[1,2*], O.S. Fatoki[3], J.P. Odendaal[2] and O.U. Oputu[3]

*Address all correspondence to: olujimiolusoji@yahoo.com or olujimio@funaab.edu.ng

1 Department of Environmental Management and Toxicology, Federal University of Technology, Alabata Road, Abeokuta, Ogun State, Nigeria

2 Department of Environmental and Occupational Studies, Cape Peninsula University of Technology, Cape Town, South Africa

3 Department of Chemistry, Cape Peninsula University of Technology, South Africa

References

[1] Fakayode, S.O. 2005. Impact assessment of industrial effluent on water quality of the receiving Alaro river in Ibadan Nigeria, AJEAM-RAGEE, 10: 1-13.

[2] Naddafi, K., Jaafarzadeh, N., Mokhtari, M., Zakizadeh, B. and Sakian, M.R. 2005 Effects of wastewater stabilization pond effluent on agricultural crops, International Journal of Environmental Science & Technology, 1 (2005) 273-277.

[3] Karamanis, D., Stamoulisc,K., Ioannides, K. and Patiris, D. 2008. Spatial and seasonal trends of natural radioactivity and heavy metals in river waters of Epirus, Macedonia and Thessalia, Desalination, 224: 250–260.

[4] Daso, A.P., Fatoki, O.S., Odendaal, J.P. and Olujimi, O.O. 2012. Occurrence of selected polybrominated diphenyl ethers and 2,2',4,4',5,5'-hexabromobiphenyl (BB-153) in sewage sludge and effluent samples of a wastewater treatment plant in Cape Town, South Africa, Archives of Environmental Toxicology and Contamination, 62: 391-402.

[5] Olujimi, O.O., Fatoki, O.S., Odendaal, J.P. and Daso, A.P. 2012. Chemical monitoring and temporal variation in levels of endocrine disrupting chemicals (phenols and phthalate esters) from selected wastewater treatment plant and freshwater systems in Cape Town, South Africa, Microchemical Journal 101: 11-23.

[6] Olujimi, O.O., Fatoki, O.S., Daso, A.P., Akinsoji, O.S., Oputu, O.U., Oluwafemi, O.S. and Songca, S.P. Levels of nonylphenol and bisphenol A in wastewater treatment plant effluent, sewage sludge and leachate from Cape Town, South Africa in "Handbook of Wastewater Treatment: Biological Methods, Technology and Environmental Impact" Edited by Cesaro J. Valdez and Enrique M. Maradona. 2013. ISBN: 978-1-62257-591-6.

[7] Akan, J.C., Abdulrahaman, F.I., Dimari, G.A. and Ogugbuaja, V.O. 2008. Physicochemical determination of pollutants in wastewater and vegetables samples along the Jakara wastewater channel in kano metropolis Kano State, Nigeria. European Journal of Scientific Research, 23: 122-133.

[8] Chen, S.L., Dzeng, S.R., Yang, M., Chiu, K., Shieh, G. and Wai. C.M. 1994. Arsenic Species in Groundwaters of the Blackfoot Disease Area, Taiwan, Environmental Science and Technology, 32: 877-881.

[9] Williams, W., Fordyce, F., Paijitprapapon, A. and Charoenchaisri, P. 1996. Arsenic contamination in surface drainage and groundwater in part of the southeast Asian tin belt, Nakhon Si Thammarat Province, southern Thailand, Environmental Geology, 27:16–33.

[10] Mukherjee, A. B. and Bhattacharya, P. 2001. Arsenic in groundwater in the Bengal Delta Plain: Slow poisoning in Bangladesh. *Environmental Reviews* 9: 189-220

[11] Jung, M.C., Thorntonb, I. and Chonc, H. 2002. Arsenic, Sb and Bi contamination of soils, plants, waters and sediments in the vicinity of the Dalsung Cu–W mine in Korea, Science of the Total Environment, 295: 81–89.

[12] Ikem, A. Egiebor, N. O. and k. Nyavor. 2003. Trace elements in water, fish and sediment from Tuskegee lake, Southeastern USA., Water Air and Soil Pollution, 149: 51–75.

[13] Iwashita, M. and Shimamura, T. 2003. Long-term variations in dissolved trace elements in the Sagami River and its tributaries (upstream area), Japan. *The Science of the Total Environment* 312: 167–179.

[14] Xia, Y. and Liu, J. 2004. An overview on chronic arsenism via drinking water in PR China, Toxicology 198 (1-3): 25–29.

[15] Gutierrez, R.., Rubio-Arias, H., Quintana, R., Ortega, J.A. and Gutierrez, M. 2008. Heavy metals in water of the San Pedro River in Chihuahua, Mexico and it potential health risk, International Journal of Environmental Research and Public Health 5:91-98.

[16] Arain, M.B., Kazi, T.G., Baig, J.A., Jamali, M.K. Afridi, H.I., Shah, A.Q. Jalbani, N., and Sarfraz, R.A. 2009. Determination of arsenic levels in lake water, sediment, and foodstuff from selected area of Sindh, Pakistan: Estimation of daily dietary intake., Food and Chemical Toxicology 47: 242-248.

[17] Zhang, Y.W., Lia, L. Huang, Y. and Cao, J. 2010. Eggshell membrane-based solid-phase extraction combined with hydride generation atomic fluorescence spectrometry for trace arsenic (V) in environmental water samples, Talanta, 80:1907–1912.

[18] Department of Water Affairs and Forestry, Water Quality Guidelines, Aquatic Ecosystem Use, 7 (1996).

[19] Canadian Council of Ministers of the Environment (CCME). 1999. Canadian water quality guidelines for the protection of aquatic life: Cadmium.

[20] Okonkwo, J.O., Arsenic status and distribution in soils at disused cattle dip in South Africa., Bulletin of Environmental Contamination and Toxicology, 79 (2007) 380-383.

[21] Fatoki, O.S., Lujiza, N. and Ogunfowokan A, O. 2002. Trace metal pollution in the Umtata River., Water SA, 28:183-190.

[22] Sanders, M.J., Du Preez, H.H. and Van Vuren, J.H.J. 1999. Monitoring of cadmium and zinc contamination in freshwater systems with the use of the freshwater river crab, Potamonautes warreni., Water SA, 25:91-98.

[23] Reinecke, A.J., Snyman, R.G. and Nel, J.A.J. 2003. Uptake and distribution of lead (Pb) and cadmium (Cd) in the freshwater Crab, Potamonautes Perlatus (Crustacea) in the Eerste River, South Africa., Water Air and Soil Pollution, 145: 395-408.

[24] Awofolu, O.R., Mbolekwa, Z., Mtshemla, V. and Fatoki, O.S. 2005. Levels of trace metals in water and sediment from Tyume river and its effects on an irrigated farmland, Water SA, 31: 87-94.

[25] Okonkwo, J.O. and Mothiba, M. 2005. Physico-chemical characteristics and pollution levels of heavy metals in the rivers in Thohoyandou, South Africa. *Journal of Hydrology* 308: 122–127.

[26] Ohimain, E.I., Jonathan, G. and Abah, S.O. 2008. Variations in heavy metal concentrations following the dredging of an oil well access canal in the Niger Delta, Advances in Biological Research, 2:97-103.

[27] Ashokkumar, S., Mayavu, P., Sampathkumar, P., Manivasagam, P. and Rajaram, G. 2009. Seasonal distribution of heavy metals in the Mullipallam creek of Muthupettai mangroves (Southeast coast of India), American-Eurasian Journal of Scientific Research 4:308-312.

[28] Bouraie, M.M., El Barbary, A.A., Yehia, M.M. and Motawea, E.A. 2010. Heavy metal concentrations in surface river water and bed sediments at Nile Delta in Egypt, Finnish Peatland Society, 61: 1–12.

[29] Papafilippaki, A.K., Kotti, M. E., and Stavroulakis, G. G. 2008. Seasonal variations in dissolved heavy metals in the Keritis River, Chania, Greece., Global NEST Journal, 10: 320–325.

[30] Li, Q. and Zhang, S. 2010. Spatial characterization of dissolved trace elements and heavy metals in the upper River (China) using multivariate statistical techniques, Journal of Hazardous Material, 176:579-588.

[31] Osman, A.G.M. and Kloas, W. 2010. Water quality and heavy metal monitoring in water, sediments, and tissues of the African catfish Clarias gariepinus (Burchell, 1822) from the river Nile, Egypt, Journal of Environmental Protection, 1: 389-400.

[32] Retief, N.R., Avenant-Oldewage, A. and Du-Preez, H.H. 2009. Water SA Seasonal study on Bothriocephalus as indicator of metal pollution in yeloowfish, South Africa, Water SA, 35: 315-322.

[33] Fatoki, O.S. and Awofolu, R. 2003. Level of Cd, Hg and Zn in some surface waters from Eastern Cape Province, South Africa. *Water SA*. 29(4): 375-380.

[34] Fernandez M.A., Gonzelez, L.M., Gonzalez, M.J. and Tabera, M.C. 1992. Organochlorine compounds and selected metals in waters and soils from Donana National Park (Spain).*Water Air and Soil Pollution*, 65:293-302.

[35] Navarro, M., Lopez, H., Sanchez, M. and Lopez M.C.1993. The effect of industrial pollution on mercury levels in water, soil, and sludge in the Coastal area of Motril, Southeast Spain, Archives of Environmental Toxicology and Contamination, 24: 11-15.

[36] Ayas, Z. and Kolankaya, D. 1996. Accumulation of some heavy metals in various environments and organisms at Goksu Delta, Turkey. *Bulletin of Environmental Contamination and Toxicology*. 56: 65-72.

[37] Ramos, L.. Fernandez, M.A., Gonzalez, L. And Hernandez, L.L. 1999. Heavy metal pollution in water, sediment and earthworms from the Ebro River, Spain. *Bulletin of Environmental Contamination and Toxicology*, 63: 305-311.

[38] Jackson, V.A., Paulse, A. N., Stormbroek, T., Odendaal, J. P. and Khan, W. 2007. Investigation into metal contamination of the Berg River, Western Cape, South Africa, Water SA, 33: 175-182.

[39] Jackson, V.A., Paulse, A.N., Odendaal, J.P. and Khan, W. 2009. Investigation into the metal contamination of the Plankenburg and Diep Rivers, Western Cape, South Africa., water SA, 35:289-299.

[40] Singh, K.P., Mohan, D., Singh, V.K. and Malik, A. 2005. Studies on distribution and fractionation of heavy metals in Gomti river sediments—a tributary of the Ganges, India Journal of Hydrology, 312(1-4):14-27.

[41] Duman, F., Aksoy, A. and Demirezen, D. 2007. Seasonal variability of heavy metals in surface sediment of Lake Sapanca, Turkey, Environmental Monitoring Assessment, 133:277–283.

[42] Marcussen, H., Dalsgaard, A. and Holm, P.E. 2008. Content, distribution and fate of 33 elements in sediments of rivers receiving wastewater in Hanoi, Vietnam., Environmental Pollution, 155:41-51.

[43] Yang, Z., Wang, Y., Shen, Z., Niu, J. and Tang, Z. 2009. Distribution and speciation of heavy metals in sediments from the mainstream, tributaries, and lakes of the Yangtze River catchment of Wuhan, China, Journal of Hazardous Materials, 166:1186-1194.

[44] Ayeni, O.O., Ndakidemi, P. A., Snyman, R. G. and Odendaal, J. P. 2010. Metal contamination of soils collected from four different sites along the lower Diep River, Cape Town, South Africa, International Journal of the Physical Sciences, 5:2045-2051.

[45] Sekabira,K., Origa, H.O., Basamba, T.A., Mutumba, G. and Kakudidi, E. 2010. Assessment of heavy metal pollution in the urban stream sediment and its tributaries, Inernational Journal of Environmental Science Technology, 7: 435-446.

[46] Bai, J., Cui, B., Chen, B., Zhang, K., Deng, W., Gao, H. and Xiao, R. 2011. Spatial distribution and ecological risk assessment of heavy metals in surface sediments from a typical plateau lake wetland, China, Ecological Modeling, 222:301-306.

Chapter 9

A Study and Analysis on the Physical Shading Effect of Water Quality Control in Constructed Wetlands

T.Y Yeh, M.H. Wu, C.Y. Cheng and Y.H Hsu

Additional information is available at the end of the chapter

http://dx.doi.org/10.5772/58892

1. Introduction

Eutrophication of water bodies has been studied since 1970. It is an urgent water pollution issue that needs to be addressed. In the past, eutrophication often occurred in reservoirs, lakes, or watersheds where water bodies have a slow flowing speed. Eutrophication or hypertophication refers to the accumulation of nutrients due to the addition of external nutrients into the water system (lakes, reservoirs, and wetlands) [7]. Eutrophication, according to nutrient concentration, can be divided into four types: oligitrophic, mesotropic, eutrophic, and hypertrophic [9].

In recent years, eutrophication in water bodies mainly was the result of excessive human activities such as overdevelopment of watersheds, agricultural irrigation, and recreation that cause the addition of exogenous nutrients to local water bodies that sped up the eutrophication of the water bodies and caused serious damage (Wu et al., 2006). The eutrophication phenomena include an increase of nutrient concentration and a rapid increase in the number of algae. However, Carlson Tropic State Index (CTSI), this is a measure of the trophic status of a body of water using several measures of water quality including: transparency or turbidity (using Secchi disk depth recordings), chlorophyll-a concentrations (algal biomass), and total phosphorus levels (usually the nutrient in shortest supply for algal growth).

Generally, the treatment technologies of eutrophic water bodies can be classified into physical, chemical, and eco-technology. Commonly used technologies include aeration at the bottom layer to increase dissolved oxygen concentrations in the deeper layers, addition of an algae inhibitor called copper sulfate that inhibits the growth of algae by increasing the concentration of Cu^{2+}, or the construction of wetlands that remove the carbon source and nutrients in the water [3, 7]. Among them, physical treatment requires huge amounts of energy to be effective and the addition of chemical agents increases treatment costs. Therefore, eco-technology treatment, a green technology, has gained high public support and at the same time, it requires lower amounts of energy, lower costs, and easy technological operations.

Eutrophication includes an increase in nutrient concentration as well as an increase in the number of algae (algal bloom). The extensive results of eutrophication are the death of a large number of animals, plants, and plankton, dramatic changes in dissolved oxygen concentration, changes in the eco-system and foul odors in the water bodies. Thus, if the nutrient concentration in water bodies can be controlled or algal bloom can be avoided, issues caused by eutrophication can be effectively rendered to prevent water quality and eco-system problems and to further achieve the purpose of water quality control.

Treatments for eutrophication mostly aim to remove the nutrients, such as nitrogen and phosphorus, in the water in order to improve eutrophication in water bodies. The mechanism of a constructed wetland system to treat eutrophication lies in the absorption of nitrogen and phosphorus nutrients by aquatic plants in the wetland system as well as the competition for the absorption of nutrients between aquatic plants and algae in water in order to control algae concentrations in water effectively. Algae in water, however, can use photosynthesis to convert light energy into chemical energy for self growth and aquatic plants can prevent the sun's ray from entering the water bodies as they grow, thereby creating the physical shading effect [5].

2. Experimental materials and methods

This study utilized model basins (as shown in Fig. 1) and in the beginning, wetland plants were collected from the wild for cultivation. Tests were started after the adaptation period of the plants. The basin size is in the dimension of L*W*H:50*35*27 cm and for each batch test, a point in each basin was used. Each day samples were collected four times; water resources came from the ecological pond of the National University of Kaohsiung. The initial nutrient concentration was also measured.

Figure 1. Module Basin Tests and Physical Shading Effect

To conduct the physical shading test, this study fixed an opaque board on the lateral side of the basins to prevent exposure to sunrays. Shading rates refers to the surface area shaded and were set as 0, 30, 50, 70, and 100%; the 100% shading rate, based on oxygen transmission, was divided into two types: sealed shading and vented sealing. Due to the shading effect developed by the different growth types of plants and the oxygen transmission to water bodies by aquatic plants, vented sealing was selected to simulate the shading effect of wetland plants.

The test interval lasted for five days and there were six points where we collected samples of shading rate every day at three time periods, every four hours from 9:00 in the morning to 5:00 in the afternoon.

Water quality parameters are used mainly for sample analysis including DO, pH, ORP, temperature, electric conductivity, and luminance, which were measured on site as well as nitrogen and phosphorous contents, Chlorophyll a, turbidity, SS, transparency, which were analyzed at the lab.

3. Results and discussion

3.1. Influence of different shading rates on dissolved oxygen concentration

Table 1 shows daily dissolved oxygen concentration at different shading rates. On Day 1 after the model basin system was built, the initial concentration level of dissolved oxygen for each group was between 5-7 mg/L while the final concentration of dissolved oxygen of the 100 (aerobic), 100 (anaerobic), 70, 50, 30, and 0% groups was, respectively, 12.00±1.26, 8.08±0.10, 22.00±0.00, 13.28±0.18, 14.77±0.36 and 14.36±0.38 mg/L. Among them, the 70, 50, 30, and 0% groups showed a significant increase of dissolved oxygen concentration.

Testing time	100% (aerobic)	100%(anaerobic)	70%	50%	30%	0%
Day 1	6.51±0.17	5.91±0.08	6.57±0.24	6.12±0.42	6.43±0.37	6.73±0.37
Day 2	7.38±0.17	6.66±0.1	8.25±0.52	8.58±0.97	8.83±1.04	9.62±1.6
Day 3	8.11±0.32	6.94±0.07	10.79±1.34	13.28±2.85	14.45±3.43	17.04±4.53
Day 5	9.22±1.24	8.54±1.23	18.8±3.74	21.71±0.41	22±0.00	22±0.00
Day 7	12.00±1.26	8.08±0.10	22±0.00	13.28±0.18	14.77±0.36	14.36±0.38

Table 1. Average Daily Dissolved Oxygen Concentration Levels at Different Shading Rates (mg/L)

Figure 2 indicates the trend of dissolved oxygen concentration at different shading rates where the 70, 50, 30 and 0% groups had significantly increased concentration levels from Day 2 to Day 5; the 50, 30, and 0% groups had decreased concentration levels from Day 5 to Day 7. The 70% group had an increased concentration level. For the 100% (aerobic) and 100% (anaerobic) groups, dissolved oxygen concentration increased during the tests but they were small changes.

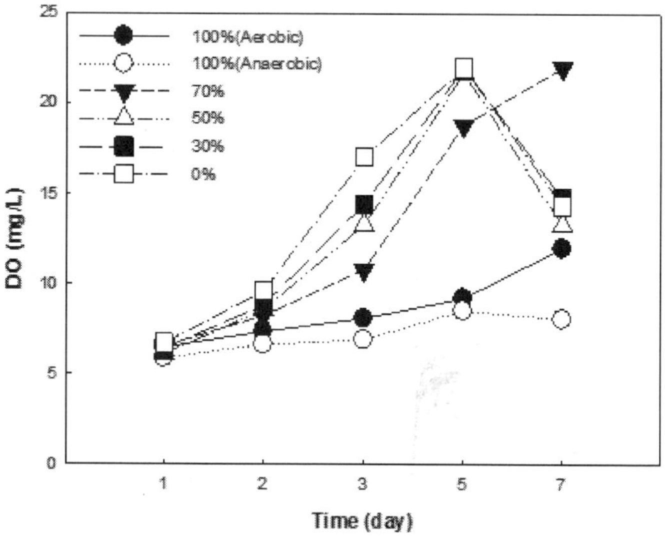

Figure 2. Dissolved Oxygen Concentration Trends at Different Shading Rates

Overall, the 70, 50, 30, and 0% groups were influenced by the photosynthesis of algae, resulting in a more significant change of dissolved oxygen concentration in water bodies. The concentration level decreases along with the increase in shading rate. For the 50, 30, and 0% groups, due to the reduced concentration level of algae, dissolved oxygen concentration decreased at the latter stage.

3.2. Influence of different shading rates on algae concentration

Table 2 shows daily algae concentration at different shading rates. As shown in the table, the initial algae concentrations of the 100 (aerobic), 100 (anaerobic), 70, 50, 30 and 0% groups were, respectively, 5.27±0.93, 5.92±0.00, 7.90 ±5.59, 5.92±0.00, 7.90±5.59, and 13.82±2.79 μg/L while the final algae concentrations of 100 (anaerobic), 100 (anaerobic), 70, 50, 30 and 0% were, respectively, 140.86±58.94, 15.8±2.79, 15.8±2.79, 53.32±4.84, 55.29±2.79 and 49.37±26.64 μg/L. For the 70, 50, 30 and 0% groups, from Day 1 to Day 7, algae concentration significantly increased.

Figure 3 shows the algae concentration trend at different shading rates in water bodies. As shown in this figure, the 70, 50, 30 and 0% groups from Day 2 to Day 5 had significantly increased concentration levels while between Day 5 and Day 7, they decreased. For the 100% (anaerobic) and 100% (anaerobic) groups, due to the shading effect, algae could not use photosynthesis and therefore, the concentration was lower.

Testing time	100%(aerobic)	100%(anaerobic)	70%	50%	30%	0%
Day 1	5.27±0.93	5.92±0	7.9±5.59	5.92±0	7.9±5.59	13.82±2.79
Day 2	1.97±2.79	15.14±9.17	19.75±7.39	15.8±7.39	21.72±12.17	25.67±21.81
Day 3	19.75±5.59	7.9±2.79	49.37±14.78	102.68±43.62	130.33±58.04	152.05±60.48
Day 5	57.27±11.17	9.87±2.79	244.86±50.34	280.4±32.92	329.77±15.55	250.78±32.21
Day 7	140.86±58.94	15.8±2.79	177.72±17.44	53.32±4.84	55.29±2.79	49.37±26.64

Table 2. Daily Algae Concentration Levels at Different Shading Rates in Water Bodies (μg/L)

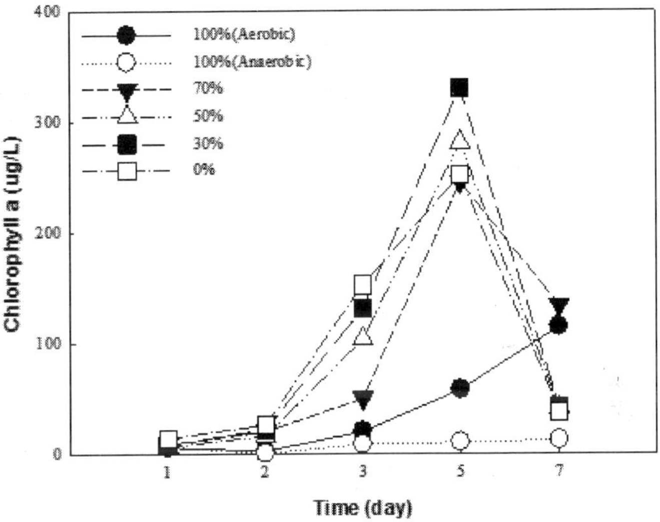

Figure 3. Algae Concentration Trends at Different Shading Rates in Water Bodies

3.3. Influence of nutrients at different shading rates

Tables 3-5, respectively, show the daily concentration of ammonia nitrogen, nitrate nitrogen, and nitrite nitrogen. Results indicate that the initial concentration levels of ammonia nitrogen for the 100 (aerobic), 100 (anaerobic), 70, 50, 30 and 0% groups were, respectively, 3.07±0.68, 2.69±0.24, 2.64±0.32, 2.69±0.41, 2.5±0.16 and 2.52±0.23 mg/L while the final concentration levels of ammonia nitrogen of the 100 (aerobic), 100 (anaerobic), 70, 50, 30 and 0% group were, respectively, 0.20±0.08, 0.40±0.08, 0.10±0.03, 0.08±0.04, 0.29±0.23, and 0.08±0.06 mg/L. Each group on Day 7 had significantly reduced concentrations of ammonia nitrogen.

Testing Time	100%(aerobic)	100%(anaerobic)	70%	50%	30%	0%
Day 1	3.07±0.68	2.69±0.24	2.64±0.32	2.69±0.41	2.5±0.16	2.52±0.23
Day 2	3.22±0.66	2.81±0.08	2.86±0.3	2.34±0.26	2.69±0.1	2.61±0.25
Day 3	0.44±0.33	0.26±0.08	0.06±0.04	0.23±0.22	0.03±0.05	0.1±0.05
Day 5	0.07±0.05	0.04±0.03	0.07±0.05	0.04±0.06	0.34±0.26	0.04±0.06
Day 7	0.2±0.08	0.4±0.08	0.1±0.03	0.08±0.04	0.29±0.23	0.08±0.06

Table 3. Daily Ammonia Nitrogen Concentration Levels at Different Shading Rates in Water Bodies (mg/L)

Testing time	100%(aerobic)	100%(anaerobic)	70%	50%	30%	0%
Day 1	1.07±0.05	1.19±0.06	1.21±0.06	1.16±0.04	1.14±0.12	1.36±0.07
Day 2	1.27±0.12	1.32±0.04	1.32±0.14	1.12±0.17	1.24±0.07	1.37±0.03
Day 3	0.9±0.08	1.07±0.03	0.87±0.05	0.87±0.17	0.76±0.04	1.01±0.34
Day 5	0.96±0.15	1.14±0.18	0.51±0.06	0.22±0.03	0.23±0.05	0.29±0.08
Day 7	0.86±0.15	1.24±0.18	0.41±0.06	0.32±0.03	0.33±0.05	0.39±0.08

Table 4. Daily Nitrate Nitrogen Concentration Levels at Different Shading Rates in Water Bodies (mg/L)

Testing time	100%(aerobic)	100%(anaerobic)	70%	50%	30%	0%
Day 1	0.09±0	0.09±0	0.08±0	0.08±0	0.08±0	0.08±0
Day 2	0.13±0.01	0.14±0.02	0.1±0	0.07±0.02	0.09±0	0.09±0
Day 3	0.18±0.01	0.25±0.02	0.11±0	0.09±0	0.09±0	0.08±0
Day 5	0.23±0.00	0.34±0.00	0.1±0	0.01±0	0.01±0	0.01±0
Day 7	0.24±0.00	0.35±0.00	0.08±0	0.01±0	0.01±0	0.01±0

Table 5. Daily Nitrite Nitrogen Levels at Different Shading Rates in Water Bodies (mg/L)

The initial concentration levels of nitrate nitrogen of the 100 (aerobic), 100 (anaerobic), 70, 50, 30 and 0% groups were, respectively, 1.07±0.05, 1.19±0.06, 1.21±0.06, 1.16±0.04, 1.14±0.12, and 1.36±0.07 mg/L while the final nitrate nitrogen concentration levels of the 100 (aerobic) and 100 (anaerobic), 70, 50, 30 and 0% groups were, respectively, 0.86±0.15, 1.24±0.18, 0.41±0.06, 0.32±0.03, 0.33±0.05, and 0.39±0.08 mg/L. On Day 7, the 70, 50, 30, and 0% groups did not have significantly reduced concentration levels of nitrate nitrogen.

http://dx.doi.org/10.5772/58892

Testing time	100%(aerobic)	100%(anaerobic)	70%	50%	30%	0%
Day 1	0.19±0.01	0.19±0.01	0.22±0.04	0.19±0.00	0.19±0.00	0.18±0.03
Day 2	0.19±0.00	0.19±0.00	0.18±0.00	0.18±0.01	0.18±0.00	0.18±0.01
Day 3	0.18±0.00	0.18±0.00	0.15±0.01	0.12±0.01	0.12±0.01	0.1±0.02
Day 5	0.16±0.01	0.18±0.00	0.05±0.01	0.02±0.00	0.02±0.01	0.03±0.00
Day 7	0.14±0.01	0.18±0.00	0.01±0.00	0.01±0.00	0.01±0.00	0.02±0.00

Table 6. Daily Orthophosphate Concentration Trends at Different Shading Rates in Water Bodies (mg/L)

The initial concentration levels of nitrite nitrogen for the 100 (aerobic), 100 (anaerobic), 70, 50, 30 and 0% groups were, respectively, 0.09±0.00, 0.09±0.00, 0.08±0.00, 0.08±0.00, 0.08±0.00 and 0.08±0.00 mg/L while the final concentration levels of nitrite nitrogen of the 100 (aerobic), 100 (anaerobic), 70, 50, 30 and 0% groups were, respectively, 0.24±0.00, 0.35±0.00, 0.08±0.00, 0.01±0.00, 0.01±0.00 and 0.01±0.00 mg/L. On day 7, the concentration levels of nitrite nitrogen for the 100% (aerobic) and 100% (anaerobic) groups increased while that of the 70, 50, 30 and 0% groups decreased.

Testing time	100%(aerobic)	100%(anaerobic)	70%	50%	30%	0%
Day 1	0.23±0.02	0.22±0.01	0.24±0.03	0.22±0.01	0.21±0.01	0.23±0.00
Day 2	0.18±0.00	0.18±0.00	0.18±0.01	0.17±0.01	0.18±0.01	0.18±0.01
Day 3	0.2±0.00	0.2±0.00	0.18±0.01	0.15±0.01	0.16±0.01	0.15±0.01
Day 5	0.18±0.01	0.2±0.01	0.1±0.01	0.04±0.01	0.05±0.01	0.05±0.00
Day 7	0.19±0.11	0.25±0.09	0.03±0.00	0.03±0.00	0.02±0.00	0.04±0.00

Table 7. Daily Total Phosphorous Contents at Different Shading Rates in Water Bodies (mg/L)

Figures 4 to 6 show the concentration trends of ammonia nitrogen, nitrate nitrogen, and nitrite nitrogen at different shading rates. Results indicate that for each group the ammonia nitrogen concentration significantly decreased on Days 2 and 3 and between Days 3 and 7, there was no significant change.

The nitrate nitrogen concentration levels of the 100% (aerobic) and 100% (anaerobic) groups decreased between Day 1 and Day 3 and from Day 3 to Day 7, they increased. The levels in the 70, 50, 30 and 0% groups decreased from Day 2 to Day 5. However, the levels for the 50, 30 and 0% groups increased between Day 5 and Day 7. The 70% group showed a continuous decrease.

The concentration levels of nitrite nitrogen, for the 100% (aerobic) and 100% (anaerobic) groups, showed an increasing trend during the tests while that of the 70, 50, 30 and 0% groups decreased along with time.

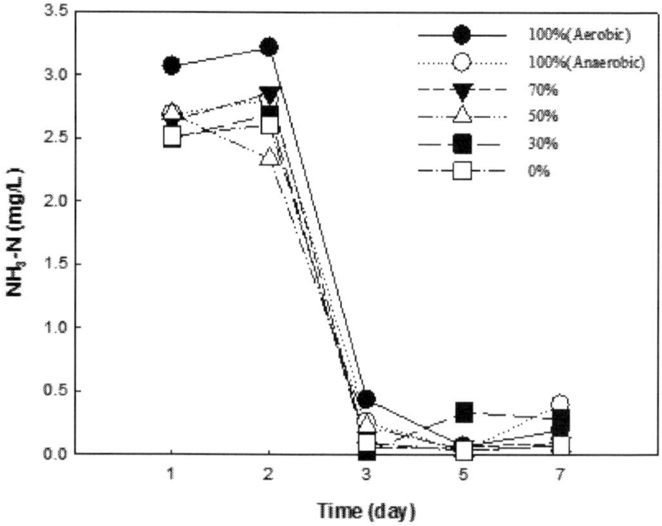

Figure 4. Ammonia Nitrogen Concentration Trends at Different Shading Rates in Water Bodies

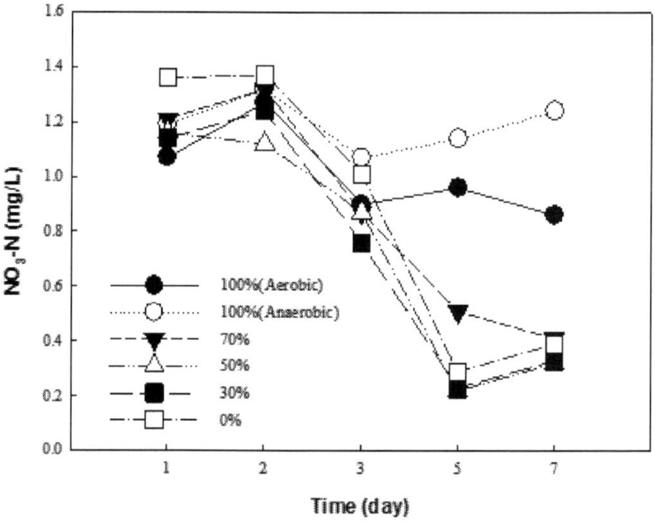

Figure 5. Nitrate Nitrogen Concentration Trends at Different Shading Rates in Water Bodies

http://dx.doi.org/10.5772/58892

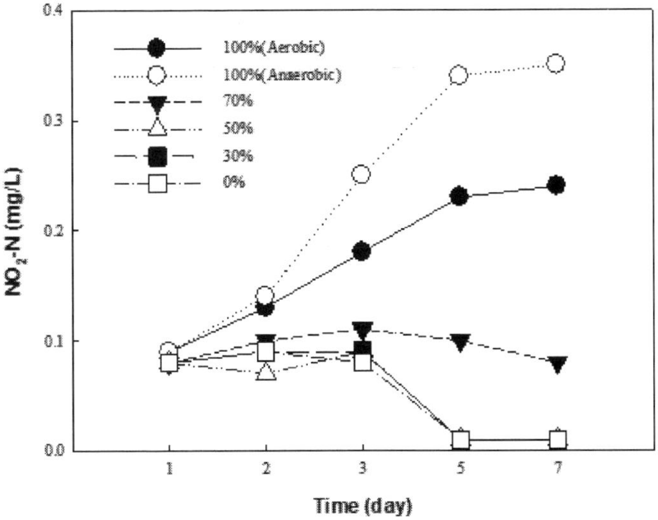

Figure 6. Nitrite Nitrogen Concentration Trends at Different Shading Rates in Water Bodies

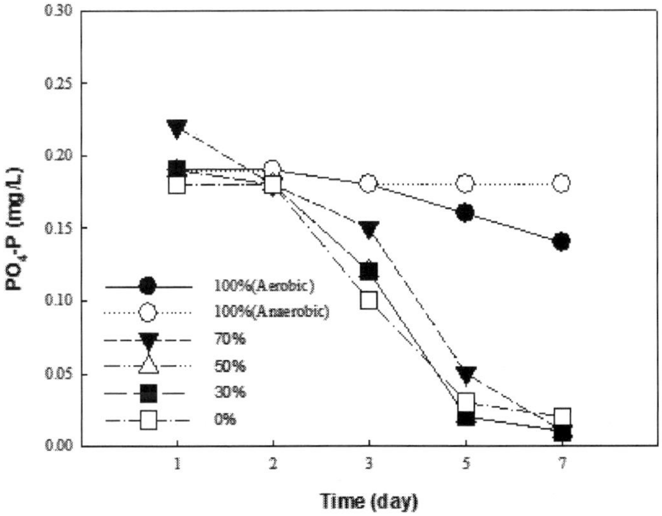

Figure 7. Orthophosphate Concentration Trends at Different Shading Rates in Water Bodies

Figure 7 and Fig. 8 show daily orthophosphate concentration and total phosphorous contents at different shading rates in water bodies. The results indicate that the initial orthophosphate concentration of the 100 (aerobic), 100 (anaerobic), 70, 50, 30 and 0% groups were, respectively, 0.19±0.01, 0.19±0.01, 0.22±0.04, 0.19±0.00, 0.19±0.00, and 0.18±0.03 mg/L while the final concentration levels of the 100 (aerobic), 100 (anaerobic), 70, 50, 30 and 0% groups were, respectively, 0.14±0.01, 0.18±0.00, 0.01±0.00, 0.01±0.00, 0.01±0.00 and 0.02±0.00 mg/L. For the 70, 50, 30, and 0% groups, there were significantly reduced concentrations from Day 1 to Day 7.

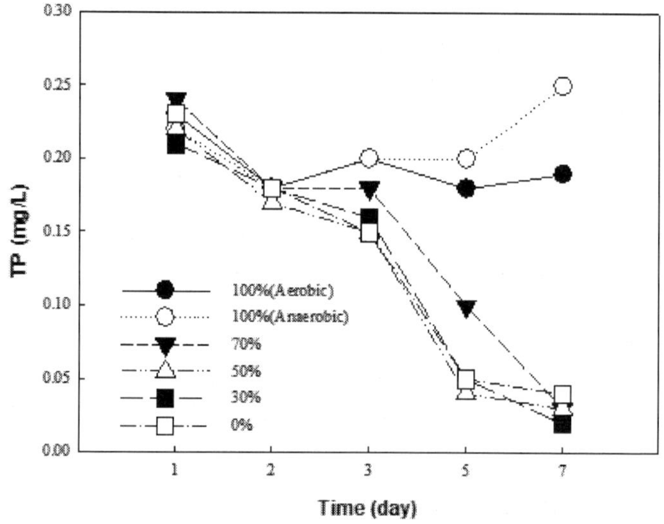

Figure 8. Total Phosphorous Content Trends at Different Shading Rates in Water Bodies

The initial total phosphorous contents of the 100 (aerobic), 100 (anaerobic), 70, 50, 30 and 0% were, respectively, 0.23±0.02, 0.22±0.01, 0.24±0.03, 0.22±0.01, 0.21±0.01 and 0.23±0 mg/L and the final total phosphorous contents of the 100 (aerobic), 100 (anaerobic), 70, 50, 30 and 0% groups were respectively 0.19±0.11, 0.25±0.09, 0.03±0.00, 0.03±0.00, 0.02±0.00 and 0.04±0.00 mg/L. The 70, 50, 30 and 0% groups, from Day 1 to Day 7, had significantly reduced total phosphorous content.

Figure 7 and 8, respectively, show the orthophosphate concentration trends and total phosphorous contents at different shading rates in water bodies. The results indicate that for each group during the tests, the orthophosphate concentration gradually decreased and among them, the 70, 50, 30, and 0% groups showed the most significant changes of orthophosphate concentrations while the 70, 50, 30 and 0% groups showed the most significant reduction of orthophosphate concentrations during the tests.

In general, due to the rapid increase in the number of algae, the 70, 50, 30 and 0% groups were found to have a significant reduction in orthophosphate concentrations and total phosphorous content. Higher shading rates resulted in higher orthophosphate concentrations and total phosphorous content.

4. Conclusion

As shown in the results of the physical shading rates, along with an increase in shading rates, dissolved oxygen concentration decreases. When the shading rate reaches 100%, there is the smallest change in dissolved oxygen concentration and water temperature decreases along with the increase of shading ratios. Algae concentration in water bodies decreases significantly along with the increase in water temperature and on Day 5, water bodies that are partially shaded has algae concentrations five times higher than that of the completely shaded water bodies. Results indicate that it is important to control algae growth in water bodies by blocking sunrays to prevent algae from having the energy, through photosynthesis, to rapidly grow in number.

There was relatively no correlation between changes in ammonia nitrogen concentrations and the physical shading effect. Nitrate nitrogen was found on Day 3 to have increased in concentration along with the increase of the physical shading rate; nitrate nitrogen was found on Day 2 to have increased in concentration along with an increase in shading rate. Orthophosphate concentrations were discovered on Day 3 to increase along with the shading ratio and total phosphorous contents were found on Day 3 to increase according to shading rates. Overall, water bodies completely shaded on Day 7 had concentration levels of nitrate nitrogen and nitrate nitrogen two times higher than that of partially shaded water bodies and seven times higher orthophosphate concentrations and phosphorous contents than that of partially shaded water bodies.

In conclusion, the above two points find that the use of the physical shading effect prevents algae from using photosynthesis to convert the light energy that is used for self growth and proliferation and the resulting rapid growth in the number of algae. The physical shading effect, however, cannot reduce nutrients in order to provide beneficial functions and that results in high nutrient levels in water bodies in the latter part of the test. After the introduction of the Carlson Tropic State Index (CTSI), the completely shaded water bodies at the initial stage of the test are mesotrophic water bodies that had neutral nutrients and partially shaded ones are eutrophic water bodies. In the latter stage of the test, the former turned into eutrophic water bodies while the latter turned into mesotrophic water bodies. The main cause lies in the relatively higher weight that CTSI places on total phosphorous content.

Author details

T.Y Yeh*, M.H. Wu, C.Y. Cheng and Y.H Hsu

*Address all correspondence to: tyyeh@nuk.edu.tw

Department of Civil and Environmental Engineering, National University of Kaohsiung, Taiwan

References

[1] Alvarez, J.A., Becares, E. 2006. Seasonal decomposition of Typha latifolia in a free-water surface constructed wetland. Ecol. Eng., 28, 99-105.

[2] Chen, Y. C. 2005. Wetland Eco-Engineering. Tsang Hi Book Publishing. Taipei, Taiwan.

[3] Cooke, G.D., Welch, E.B., Peterson, S.A., Newroth, P.R. 1993. Restoration and Management of Lakes and Reservoir. 2nd, ed., Lewis Publishers, Boca Raton.

[4] Dodds, W.K., Jones, J.R., Welch, E.B. 1998. Suggested classification of stream trophic state: distributions of temperate stream types by chlorophyll, total nitrogen, and phosphorus. Water Res., 32, 1455-1462.

[5] Ke, D. Y. 2010. A Study and Analysis of Water Quality Improvement by Controlling Algae Growth in Natural Purification Water System, MA Thesis, Department of Civil and Environmental Engineering, National University of Kaohsiung.

[6] Kadlec, R.H., Knight, R.L., 1996. Treatment Wetlands, CRC Press, USA.

[7] Kuo, C. T., Wu, C. T., Wu, H. C., Lin, C. F., Lung, W. S., Wu, M. C., Chen, Y. C., Yang, C.B. 2005. A Research Project on Eutrophication Control by Purifying Water Quality at Reservoirs with Eco-Engineering Method. Environmental Protection Administration, Executive Yuan, the Republic of China.

[8] Lin, Y. K. 2010. A Wetland Policy Study based on Managerial Concept of Eco-system. M.A. Thesis. Institute of Natural Resources Management, National Taipei University.

[9] Nurnberg, G., 1996. Trophic state of clear and colored, soft- and hard-water lakes with special consideration of nutrients, anoxia, phytoplankton and fish. Lake and Reservoir Management, 12, 432-447.

Chapter 10

Economic Feasibility Study of Canal Plastic Lining in the Aral Sea Basin

Inna Rudenko, Sanaatbek Salaev,
Sanjar Davletov and Ruzumboy Eshchanov

Additional information is available at the end of the chapter

http://dx.doi.org/10.5772/59101

1. Introduction

The Aral Sea Basin (ASB) in Central Asia centuries ago was part of the Great Silk Road that brought Chinese silk, bronze ware, cosmetics, paint, rice, and tea to the West, and glassware, dried fruits, vegetables, cotton, horses, and semi-precious stones to the East [5]. However, the ASB was a place not only for trade but also a cradle for cultures, ideas and agricultural development, a place where impressive irrigation and drainage networks have been established to support the flourishing of the oasis.

Since the ancient days agriculture in the ASB region has been possible only with irrigation but viable for millions of farm families to make a living and to have access to sufficient drinking water and healthy food. Since the 1960s, about 8 million ha of land, including natural forest and desert areas, were transferred to irrigated agricultural production in the ASB. The required ca. 96 km^3 of irrigation water was conveyed through 323 000 km of channels [6]. In the past four decades, between 80 to 95% of water from the Amudarya and the Syrdarya rivers – the main feeding rivers of the Aral Sea – with annual flows of around 75 km^3 and 34 km^3 respectively, has been used for irrigation purposes for production of cotton, rice and other crops. However, the present management of irrigated cropland is becoming increasingly unsustainable with a widespread land and water resources degradation, which further threaten the ecological and economic sustainability as well as food security and health of the population in the ASB region.

The case study region Khorezm is located between 60.05 and 61.39 N and 41.13 and 42.02 E in the northwest of Uzbekistan and ASB. Khorezm is a living habitat to over 1.7 million people (as of 2011). Roughly 260 000 ha of the region are used for irrigation purposes. Agriculture in

Khorezm is possible with irrigation only to compensate the difference between the low precipitation (annually 100 mm) and high evaporation rates (up to 1400-1600 mm) resulting from the continental climate. Irrigated agriculture in Khorezm contributes to more than 50% of the regional income, provides more than 98% of hard cash revenues, and employs more than 60% of the economically active population [7].

The probability of adequate water supply has been decreasing over the past years [1]. Seasonal variations in river runoff also decrease water availability during the vegetation period [1]. The mentioned water scarcity has been aggravated by external factors, such as river runoff reduction due to climate change and the growing water demand in upstream countries [2], yet also by internal factors, including the expansion of the production of water intensive crops such as cotton and rice and the poor condition of the irrigation and drainage infrastructure causing high water losses [3].

Low conveyance and irrigation efficiencies are thus the main causes of irrigation water losses in the ASB and Khorezm. Water is mainly conveyed by unlined (earthen) irrigation canals [3]. The loss of irrigation water due to seepage and evaporation out of the irrigation canals has raised groundwater levels, thereby causing water-logging of the soil, increased salt concentrations (via evaporative concentration), and lowered crop yields [8]. Additionally, the quality of drinking water, which is often pumped from groundwater, has been negatively affected by increasing salt concentrations.

It is expected that climate change and increasing water use in upstream regions of the basin will change the quantity and temporal behavior of the water resources available in Khorezm for the worse [4]. This underlines the need for improving irrigation efficiency and adequacy of water management in Khorezm, which presently are notoriously low [3]. The enormous water losses in the irrigation system and the shallow ground water table with outcoming negative consequences call for urgent solutions to improve water supply and to increase water use efficiency. Improving irrigation efficiency is urgently needed to (i) reduce the currently enormous waste of water, (ii) contribute to overcoming the underutilization of agricultural yield potentials, and (iii) lower adverse impacts on the soil and groundwater resources [4].

2. Current situation of water use in Khorezm

The Amudarya river is a muddy river and used to bring large amount of sediments with irrigation water. These sediments covered the beds of earthen irrigation channels thus lowering infiltration of water and seepage looses. However, after the construction of the Tuyamuyun water reservoir, which main function is to store water, water discharged from the reservoir to the channels became 10-20 times less muddy. As a result, the channels do not receive sediments for bed covering and infiltration of water in the channels is high, the ground water table in the vicinity of the channels is rising, causing secondary soil salinization processes.

Irrigation water management is supported by and conducted with a larger number of dams, pumping stations also for water lifting, canals (lined but mainly earthen), intake structures,

http://dx.doi.org/10.5772/59101

flumes and subsurface and vertical drains. The larger infrastructures used to be regularly maintained and operated in contrast to the smaller structures such as pumping stations and intakes from channels as well as the drains, which are insufficiently maintained [3].

Irrigation distribution network comprised of hundreds interfarm canals in Khorezm is characterized by the excessive length – each ha of irrigated agricultural land is serviced by 37.5 meters of irrigation channels. Irrigation water is conveyed to the farm fields through around 2 445 km of channels, of which only 233 (or less than 10%) have concrete lining.

Around 90% of the channels in the Khorezm region have been constructed without any lining measures and are considered earthen canals. Furthermore, the current layout and status of the irrigation and drainage infrastructure also hinder appropriate operation. The present situation is characterized by a large number of small size irrigation fields and a large number of water users with diversified requirements. On the contrary, the irrigation and drainage network has been constructed to serve large production units (kolkhozes and sovkhozes) with high uniformity [2]. With this regards the delivery performance ratio[1] of irrigation water in Khorezm is low and enormous amount of water is lost during conveyance from the water source to the field. As a result, the fields and crops do not get the required amount of irrigation water, which influences yields, income, and food provision of the farming population. Furthermore, the estimated poor irrigation efficiency (30-33%), and high drainage ratio (55%), indicate inefficient water management at irrigation network level and especially at field level. Furthermore, the widespread occurrence of deteriorated hydraulic structures and lowered discharge capacity of the canals are consequences of a low maintenance intensity [9, 3]. Loss of irrigation water is not unique to Khorezm; worldwide, 40-75% of irrigation water is lost due to evaporation and seepage [10, 11]. As global water supplies come under increased pressure due to population growth and climate change, measures to balance agriculture and the hydrologic system are critical.

3. Water saving technologies

Many measures to increase water use efficiency at different levels have been developed so far, varying from measures for improving conveyance and distribution efficiency to measures for increasing irrigation efficiency at field level. Since irrigated agriculture is the dominant livelihood form in the study region, crop production should be ensured under conditions of reduced water supply, which would necessitate upgrading irrigation networks and management practices and improving water application at field level. The present lack of maintenance of the irrigation infrastructure engraves the on-going deterioration as evidenced by the ever-growing number of silted up and damaged canals, broken gates, outdated pumps, lack of spare parts, and so on. Hence, an improvement of the irrigation infrastructure bears high potential to decrease overall water losses, although it recently was postulated that rehabilitating and

1 The relation between actual and intended amount of water directed to a scheme or part of a scheme is the most important indicator to assess the operational performance of water distribution

renovating the irrigation systems by, for instance, concrete lining of channels could reduce irrigation water losses, but would require extraordinary high investments [12].

Given the present economic and ecological situation, it is necessary to consider not only high returns to investments, as predominantly has been emphasized in the past, but also their relevance for environmental sustainability, while considering also the financial viability of the technical measures for improving conveyance, irrigation, and management efficiencies. At field level water use reduction can be achieved by implementing improved irrigation technologies and water-wise options such as laser guided land leveling, drip or sprinkler irrigation and other through which water use efficiency at field level could be increased but where the costs of technologies showed an inverse relationship [13].

At channel level there also exist various methods for reducing water loss due to seepage and infiltration including: reinforced and unreinforced concrete; geomembranes; flocculant polymers; compacted soils and clays; mud; and plastic liners [14]. These methods have their advantages and disadvantages, but have the common explicit objective to reduce water loss, to lower groundwater level and improve water quality. None of these methods have been examined for their efficiency in the irrigated systems in Central Asia and ASB. In our study we selected the plastic lining to decrease water loss and lower groundwater levels due to appropriateness for the soil type and climate, low cost, local availability of resources, and low maintenance needs, which would potentially be to the benefit of farmers.

4. Plastic lining of channel beds for reducing infiltration

Seepage of irrigation water in the earthen canals can be reduced by placing a plastic liner on the canal bed prior to irrigation season. Lining of canals is a well-tested remediation method for minimizing seepage loss from canals.

4.1. About the technology

In 2009-2012 researchers from Urgench State University in Khorezm together with one of the Water Consumers Associations in Khorezm have tested the technology of plastic lining on one of the small interfarm earthen channels in the region.

The selected channel, 2.6 km long, provided irrigation water to 400 ha of farm land and had the discharge capacity of 1.5-2 m^3 per second. The average performance ratio hardly reached 0.49, meaning that 51% of water in the channel was lost due to seepage and infiltration.

When applying the technology an area slightly larger than the banks and bottom of the canal was cleaned and excavated of soil. Prior to excavation works the channel was checked for elevation with the use of special leveling laser tools as to ensure gravity flow of water after plastic lining. Approximately 10 to 15 cm of sand was placed as a base on the cleaned bottom of the channel (refer to the photos). The plastic was laid on top of the sand. The plastic was then covered by up to 0.5 m layer of compacted soil to keep the plastic in place, and prevent exposure to the sun.

http://dx.doi.org/10.5772/59101

Figure 1. Appearance of the channel prior to plastic lining

Figure 2. Cleaning the bottom and the banks of the channel

Figure 3. Lining of plastic

Figure 4. Covering of plastic with compacted soil

Specifications of the plastic liner included 100 μm thickness, 7 m wide, and elasticity of 250%. Such plastic is locally available and is usually and widely used by the rural population to cover the temporary greenhouses for year round production of vegetables and green vegetables. The

Figure 5. Gravity flow of water in the plastic lined channel

plastic laid at the bottom of the channel can last up to 50 years given the proper coverage by compacted soil and non-exposure to the UV sun rays.

Concurrent with application of plastic liner, fields irrigated from the selected channel were lazer leveled to improve gravity flow. The combination of enhanced gravity flow and water saved resulted in less use of mechanized pumps.

4.2. Pros and cons of the technology

The major achievement of the technology was that average performance ratio in the lined channel increased from 50 to 89% during the growing season. Overall the findings illustrate that plastic lining was effective at reducing water loss due to seepage, groundwater levels decreased in the plastic-lined canal. This was the intended effect of the plastic liner. However, though the plastic lining remained effective at reducing water seepage, as seen in the consistently lower groundwater level, it could not counteract movement of groundwater from nearby areas. High hydraulic conductivity values allowed groundwater from nearby irrigation to seep into the area around the canal thus increasing groundwater levels, which is a common, annually observed phenomena in the entire region [8]. The reasons may include the absence or low efficiency of particular crops contributing biodrainage potential in the irrigated area [15].

Reduced seepage or infiltration of irrigation water allowed saving enormous amount of water (over 10 million m^3 during one irrigation season for the total length of 2.6 km), which was used to irrigate additional cropping areas amounting to about 500 ha in the tail end of the channel. Thus, the results showed that given the same amount of water in the channel and given virtually no seepage losses farmers can irrigated twice as large cropping areas, can get additional revenues from additional cropping areas, but also from higher yields (i.e. cotton yield was at least 15% higher compared to other fields) caused by the timely delivery of irrigation water and lower ground water level.

Another major achievement was the gravity flow of water in the lined channel due to leveling measures. Gravity flow allowed the farmers to stop using pumps for delivering water both to the tail end of the channel and to the fields with higher elevation levels compared to the channel. Thus, the farmers could save enormous financial resources for electricity or diesel, not talking about the benefits to the environment.

Water shortage often happens during the peak irrigation season negatively affecting crop yields. Plastic-lined channels hold potential to be used also as decentralized water reservoirs for tackling the problem of temporarily unreliable water supplies. Filling the reservoirs (lined channels) during periods of oversupply of irrigation water could make the farmers more independent from the water supply by the network.

Along with all the evident benefits of plastic lining as an efficient water saving option, there are some obstacles for further dissemination in the region. This is in the first place the financial aspect or high costs of plastic lining (over 15 thousand USD per 1 km of channel). Farmers in the ASB are predominantly not rich and could not afford adopting this option individually. Broad adoption of this technology would require the joint efforts of many farmers using the same channel for irrigating their fields. However, interested in this water saving technology are farmers in the tail end of the channel, whereas farmers in the head of the channel receive water in any case and do not usually want to cooperate and to invest in the technology to the benefit of other farmers. Thus not farmers, but higher level structures, such as Water Consumers Associations, local governments, etc. and decision makers should become the major driving force for disseminating water saving technologies in the environmentally degraded ASB region.

4.3. Cost benefit analysis

Any innovation, any technology to be accepted and adopted by local farmers and decision makers on higher level should be financially analyzed to show the costs, the benefits and payoff period of the required investments. Economists from Urgench State University have conducted cost benefit analysis of the plastic lining technology both per 1 km of interfarm channel (rather small channel with 5-7 m width) and for the whole Khorezm region.

The technology of plastic lining required 15.3 thousand USD per 1 km of channel length (Table 1). Since it was possible to save around 4 million m^3 of irrigation water per 1 km of channel and given the cost of water delivery of around 0.002 USD per m^3 the farmers could save around 9.5 thousand USD on water delivery. Furthermore, due to gravity flow in the lined channel, the farmers could save around 3 thousand USD on energy costs for pumps and around 0.5 thousand USD on pumps maintenance.

The timely and in required amount delivery of water to the fields resulted in an increase in particular of cotton yields of up to 0.4 tons per ha. Cotton was used in analysis since it is the main cash crop cultivated in the region. If we assume that at least half of the cropping area supplied by the channel is covered with cotton, then additional income from increased cotton yield could reach 35.6 thousand USD (Table 1).

The efficiency or profitability of plastic lining technology can be analyzed from two perspectives: (1) on farmers' level (without the consideration of benefits from water delivery costs saving) and (2) on Water Consumers Association level (without the consideration of benefits from increased cotton yields). The farmers thus could receive up to 39 thousand USD of net benefit in the first year if accounted for cotton or 3.6 thousand USD not accounting for cotton revenues. The payoff period for farmers would be 0.4 years with cotton revenues, but more than 4 years without cotton revenues. The payoff period for the Water Consumers Association would be slightly over 1 year.

	Indicators		**Per 1 km of channel**
1	PL costs, USD		15 324,4
2	Additional irrigated area, ha		276,3
3	Total irrigated area, ha		476,3
4	Canal efficiency, %		89
5	Increase in canal efficiency, %		39
6	Water saving, m^3		4 019 593,8
7	Saved water delivery costs, USD		9 570,5
8	Energy for pump, KVt		60 000,0
9	Energy saving, USD		3 205,7
10	Maintanance costs for pump, USD		476,2
11	Increased yield, t/ha		0,39
12	Extra-yield income of cotton, USD		35 622
	Farmers		
13	Total benefits from PL, USD (9+10+12)	39 304	
14	Total benefits from PL without cotton, USD (9+10)	3 682	
15	Net benefits from PL, USD (13-1)	23 979,3	
16	Net benefits from PL without cotton (14-1)	-11 642,5	
17	Cost recovery, in years (1/13)	0,4	
18	Cost recovery without cotton, in years (1/14)	4,2	
	Water Consumers Association		
19	Total benefits from PL, USD (7+9+10)	13 252	
20	Net benefits from PL (19-1)	-2 072,0	
21	Cost recovery, in years (19/1)	1,2	

PL=plastic lining

Table 1. Cost benefit analysis of plastic lining on 1 km of channel

4.4. Economic feasibility of plastic lining for the whole region

Irrigation channels in the Khorezm region were laid on soil of various types from loamy to sandy, which determines the varying delivery performance ratio (or seepage amounts) of channels across the region. The technology of plastic lining could be applied to all the in-farm and interfarm channels in the region, but due to high investment costs, it would make sense to line in the first place channels with beds (bottoms) laid on sand loamy, light loamy and sandy soils, i.e. where most infiltration of irrigation water occurs. According the GIS lab of Urgench State University, there are 1 080 km of such channels in the Khorezm region. Furthermore, if only in-farm and interfarm, small channels are selected, then plastic lining should be applied to only 557 km of channels (Table 2).

Table 2 presents the results of economic feasibility of applying plastic lining in the whole region. The analysis was conducted according to 4 scenarios: scenario 1 covers all irrigation channels; scenario 2 covers interfarm and in-farm channels laid on all soil types; scenario 3 presents calculations for interfarm and in-farm channels laid on light loamy, sandy loamy and sandy soils; and finally scenario 4 covers 40% of the channels from scenario 3 since this amount of channels has the specified width of 5-7 meters. Scenario 4 can be considered the real and first hand to do scenario in the region in order to urgently prevent high water loss due to seepage in the irrigation channels.

Current water loss from all channels in the Khorezm region due to seepage reaches 2.2 km^3 of valuable irrigation water, or about 920 thousand m^3 of irrigation water per 1 km of channel length. Based on rough calculations after plastic lining of all the channels, water loss would amount to 0.5 km^3 of water or 202 thousand m^3 of water per 1 km of channel (Table 2). Thus, water saving potential as a result of plastic lining measures in the framework of the real scenario 4 stands at 0.4 km^3 of irrigation water or about 9% of total annual water intake from the Amudarya river for irrigation purposes of the Khorezm region. This saved water instead of being infiltrated to the groundwater and thus lost, alternatively could be used to irrigate more than 22 thousand ha of agricultural lands in the region.

According to scenario 4 the region would need to invest around 8.5 million USD (Table 2), whereas the benefits in the first year after plastic lining would reach 6.5 million USD including:

- Saving on water delivery – 952 thousand USD
- Saving on energy (pumps) – up to 1.8 million USD
- Saving on pumps maintenance costs – 265 thousand USD
- Additional revenue from increased cotton yields – 3.6 million USD

If taking into account additional revenue from increased cotton yields, the payoff period for plastic lining in the whole region will come due in slightly over 1 year, or 2.8 years without cotton revenues.

Indicator	for all canals	Interfarm and in-farm channels on all soil types	Interfarm and in-farm channels on light loamy, sand-loamy, sandy soils	40% of interfarm and in-farm channels (5-7 m wide) (real scenario)
Canal length, km	2 445	2 029	1 080	557
Total intake, km^3	5	5	5	5
Conveyance losses, %	50	50	50	50
Conveyance losses, km^3	2,25	1,87	0,99	0,51
Conveyance losses per 1 km, m^3	920 328,21	920 328,21	920 328,21	920 328,21
Efficiency in canals prior to PL	0,39	0,39	0,39	0,39
Efficiency in canals after PL	0,89	0,89	0,89	0,89
Conveyance losses after PL, %	11	11	11	11
Conveyance losses after PL, m^3	0,50	0,41	0,22	0,11
Conveyance losses after PL per 1 km, m^3	202 472,21	202 472,21	202 472,21	202 472,21
Water saving, km^3 after PL	1,76	1,46	0,78	0,40
Water saving, in % to total water intake	39	32	17	9
Cost of PL, USD	37 464 705	31 086 253	16 547 203	8 538 512
Saved water delivery costs, USD	4 178 571	3 467 160	1 845 568	952 331
Energy consumption for pumps, KVt	146 686 800	121 713 035	64 787 813	33 431 119
Energy saving after PL, USD	7 837 266	6 502 954	3 461 520	1 786 177
Cost savings for maintenance of pumps, USD	1 164 181	965 976	514 189	265 326
Average area under cotton, ha	105 000	87 124	46 376	23 930
Increase in yield (15%), tons per ha	0,39	0,39	0,39	0,39
Increase in income from cotton, USD	15 707 250	13 033 055	6 937 491	3 579 810
Total benefits from PL, USD	28 887 269	23 969 145	12 758 769	6 583 644
Total benefits without cotton revenues, USD	13 180 019	10 936 090	5 821 278	3 003 834
Net benefits from PL, USD	-8 577 436	-7 117 108	-3 788 435	-1 954 868
Net benefits without cotton revenues, USD	-24 284 686	-20 150 163	-10 725 926	-5 534 678
Cost recovery, years	1,30	1,30	1,30	1,30
Cost recovery without cotton revenues, years	2,84	2,84	2,84	2,84

Table 2. Economic feasibility of plastic lining technology for the whole Khorezm region

5. Discussion and conclusions

Plastic lining was studied as a first step towards addressing in general counterbalancing lower crop yields in Khorezm due to raised groundwater tables causing soil salinisation. Soil salinization and near-surface groundwater levels are a global problem. As global water supplies come under increased pressure due to population growth and climate change, measures to balance agriculture and the hydrologic system are critical. This study contributes by providing a better understanding of the promises and limitations of a lined canal for addressing increased groundwater levels and salinity resulting from intensely irrigated agriculture. Research impacts will allow Uzbek stakeholders, such as land managers and policy officials, to better address regional water resource issues (supply and quality), as well as potential soil improvement options.

Furthermore, combining the introduction of water-saving measures as plastic lining with decentralized reservoirs would create a win-win situation by tackling both shortcomings that affect the water productivity, i.e., unreliable water supplies and irrigation timing. This should go hand in hand with raising the willingness of farmers to introduce and invest in water-saving technologies.

The results showed that plastic lining of channel beds as water saving technology is a viable, efficient and rather acceptably moderate-cost option for preventing water loss and lowering ground water in the region. However the question comes on the source for required investments. The farmers in the region do not possess enough farm capital to uptake this option themselves. Thus decision makers on higher level, such as Water Consumers Association, local governments or even environmental funds should take the leading role in disseminating this technology throughout the region.

Author details

Inna Rudenko*, Sanaatbek Salaev, Sanjar Davletov and Ruzumboy Eshchanov

*Address all correspondence to: irudenko@mail.ru

Urgench State University and NGO "KRASS", Urgench, Uzbekistan

References

[1] Müller M., (2006) A General Equilibrium Modeling Approach to Water and Land Use in Uzbekistan. (Ph.D. Thesis) Bonn University, Bonn, Germany.

[2] Djanibekov N., Bobojonov I., Lamers J.P.A. (2012) Farm reform in Uzbekistan. In: Martius C, Rudenko I, Lamers JPA, Vlek PLG (Eds.) Cotton, water, salts and soums-economic and ecological restructuring in Khorezm, Uzbekistan. Dordrecht Heidelberg London New York, ISBN 978-94-007-1962-0, pp. 95-112.

[3] Tischbein B., Awan U.K., Abdullaev I., Bobojonov I., Conrad C., Jabborov H., Forkutsa I., Ibrakhimov M., Poluasheva G. (2012) Water management in Khorezm: current situation and options for improvement (hydrological perspective). In: Martius C, Rudenko I, Lamers JPA, Vlek PLG (Eds.) Cotton, water, salts and soums-economic and ecological restructuring in Khorezm, Uzbekistan. Dordrecht Heidelberg London New York, ISBN 978-94-007-1962-0, pp. 69-92.

[4] Manschadi A.M., Oberkircher L., Tischbein B., Conrad C., Hornidge A-K., Bhaduri A., Schorcht G., Lamers J.P.A, Vlek P.L.G. (2010) "White Gold" and Aral Sea disaster – Towards more efficient use of water resources in the Khorezm region, Uzbekistan. Lohmann information, Vol. 45 (1), April 2010, pp. 34-47.

[5] Tolstov S.P. (1948) Following the tracks of ancient Khorezmian civilization. Academy of Sciences of the USSR, Moscow – Leningrad

[6] Orlovsky N., M. Glanz, and L. Orlovsky (2000) Irrigation and Land degradation in the Aral Sea Basin. Pages 115-125 in Breckle S.W., Vesle M. and Wuecherer W. (eds.) Sustainable Land Use in Deserts. Springer Verlag Heidelberg, Germany.

[7] Rudenko I., J.P.A. Lamers & U. Grote (2009) *Can Uzbek farmers get more for their cotton?*, The European Journal of Development Research, vol.21 (2), pp. 283-296

[8] Jabbarov H., R.Eshchanov, J. Nurmetov, A. Lutz. M. Shanafield, J.P.A. Lamers (2013) Canal lining to increase water use efficiency and remediate groundwater levels in Khorezm Uzbekistan, Central Asia. International Journal of Agriculture: Research and Review. Vol., 3 (4), 742-750, 2013 Available online at http://www.ecisi.com ISSN 2228-7973 ©2013 ECISI Journals

[9] Awan U.K. (2010) Coupling hydrological and irrigation schedule models for the management of surface and groundwater resources in Khorezm, Uzbekistan. Dissertation, University of Bonn.

[10] Rosegrant M., Cai W., Cline S. (2003) Will the world run dry? Global water and food security. Environment 45, 24–36.

[11] Postel S., Vickers A. (2004) Boosting Water Productivity, State of the World 2004, W.W. Norton and Company, New York.

[12] Micklin P., (2002) Water in the Aral Sea Basin of Central Asia: cause of conflict or cooperation? Euroasian Geogr. Econ. 43 (7), 505–528

[13] Bekchanov M., Lamers JPA, Martius C. (2010) Pros and cons of adopting waterwise approaches in the lower reaches of the Amu Darya: A Socio-Economic View. Water, 2(2), 200-216; on line at: doi:10.3390/w2020200.

[14] Susfalk R., Sada D., Martin C., Young M., Gates T., Rosamond C., Mihevc T., Arrowood T., Shanafield M., Epstein B, Lutz A., Woodrow J., Miller G., Smith D. (2008) Evaluation of Linear Anionic Polyacrylamide (LA-PAM) Application to Water Delivery Canals for Seepage Reduction. DRI Publication Number 41245.

[15] Khamzina A., Sommer R., Lamers J.P.A., Vlek P.L.G (2009) Transpiration and early growth of tree plantations established on degraded cropland over shallow saline groundwater table in northwest Uzbekistan. Agricultural and Forest Meteorology 149,1865–1874.

Chapter 11

Water Quality of Agricultural Drainage Systems in the Czech Republic — Options for Its Improvement

Petr Fučík, Antonín Zajíček, Renata Duffková and Tomáš Kvítek

Additional information is available at the end of the chapter

http://dx.doi.org/10.5772/59298

1. Introduction

Worldwide, artificial agricultural drainage systems create the optimal conditions for crop planting and soil cultivation by removing of excess water from the root zone and providing better trafficability for farm machinery. Although they are benefitial for agricultural production, the extensively used drainage systems considerably affect the hydrological and hydrochemical regimes of catchments in both positive and in negative ways. The runoff characteristics are some of the most affected catchment parameters: differences in runoff volumes and temporal variations of flow rates in the water courses, the lowered groundwater levels and changes in the surface energy balance have frequently been found [1-4]. Problems such as altered hydrological patterns and impaired drainage water quality in agricultural catchments have often been mentioned. Water quality from agricultural drainage systems (both tiles and ditches) has been discussed by the studies which draw attention to the reduced quality of drainage waters caused by elevated concentrations of nutrients (N, P, C) and/or pesticides. The results of direct monitoring of drainage groups or very small drained catchments [5-11] together with various model approaches have proven that the contribution of agricultural drainage to water pollution in larger areas may be significant [12-15]. In principle, land drainage increases the aeration of the soil profile and thus promotes the mineralization of soil organic matter and reduces denitrification in previously waterlogged soils. Further, the systems are often connected to soil preferential flow paths and so contribute to a rapid movement of water and soluble or particle-bound contaminats to related water bodies [6,10,11,13].

In the Czech Republic (CR), there were more than 1 078 000 ha of land drainage built by 1990, which cover about ¼ of the agricultural land in the CR [16]. Many of these systems are located in slopes. Tile drainage systems, built in slopy areas underlaid by crystalline bedrocks, receive in most cases not only the water infiltrating directly from upper soil horizons, but also water rising up from the groundwater table. In these cases tile drainage is often connected to a spring, a shallow aquifer or to a groundwater body, with a broader contributive area – a drainage subcatchment, see e.g. [12,17]. In such cases, this entire drainage subcatchment must be taken into an account for water balance studies, since a considerable portion of drainage runoff originates outside the drained area [17-19]. In the conditions of hilly landscape and soil environment located upon the rocks of crystalline complex, approx. 40% of the total catchment discharge happens as subsurface flow and 30% as groundwater flow [18]. Within a drainage subcatchment, different catchment zones exhibit different hydrological and hydrochemical roles and responses, influenced predominantly by soil conditions and land use and management rather than by the technical characteristics of the tile drainage-drainage spacing and lodgment, drain diameter, etc. [20,21].

Generally, from a hydrogeological viewpoint, a catchment can be divided into recharge zones, where precipitation infiltrates quickly and then recharges the groundwater body, and discharge zones, where groundwater approaches the land surface or a surface water body [22]. The recharge zones are mainly located in the uppermost areas of the catchment, close to the catchment divide, peaks and ridges. The soils of these zones are typically shallow and stony, with high sand content and high infiltration capacity. The coarse-textured soils of the recharge zones are, with respect to groundwater resources, well suited to growing grass, which, besides benefiting water quality, increases their field capacity and results in virtually complete infiltration of precipitation, including rainstorms [2,18,23]. The recharge zones are assumed to be the most important agent in drainage runoff and water quality formation. Discharge zones can usually be found in the lowest parts of the slopes (also along water courses, lakes) and are prone to surface waterlogging. The dominant soils in the discharge zones are generally deep, with higher clay content and a lower capacity for infiltration. The recharge zones and discharge zones are connected by transient zones, where precipitation is mostly transformed into surface runoff and groundwater flows downslope in a quasi-steady way [18,24]. The transient zones are located mainly in the middle sections of slopes. Groundwater in natural catchments flows from the recharge zones to the discharge zones. The actual spatial distribution of these zones depends on the local geologic and geomorphologic conditions [25, 26, © [2011] - 27]. The aforementioned zones, being described as recharge zones, are delineated by various methods and techniques [28-33]. However, the most common is the *DRASTIC* method [34]. Other methods are e.g. *SINTACS*, which is a modified Drastic approach published by [35], or the *PI* [36] and *COP* [37] methods. In the Czech Republic, the system of soil vulnerability evaluation based on the soil classification system was developed by [38], and later improved by [39], who transformed it into the Method of Critical Source Areas Identification for the Evaluation of Agricultural Non-Point Pollution Sources. Another useful tool is the Synthetic Map of Groundwater Vulnerability Assessment published by [40].

Tile drainage systems represent a shortcut between recharge and discharge zones which significantly reduces catchment water residence time [11,41], hastens the precipitation-runoff reaction, and shortens the time to reach peak discharge during events [4]. By ploughing shallow soils of recharge zones, nitrate concentration increase in soil water, especially during the non-growing season, due to the elevated mineralization of soil organic matter, encouraged by the aeration of soil profile induced by tilling. Subsequent enhanced leaching of nitrates occurs and lasts as long as the drainage system functions [14,42-43]. These facts have led to the assumption, that land use and agricultural management of areas prone to rapid infiltration profoundly affect nitrate concentration in surface as well as groundwaters [31,44-46].

In this chapter, results from two experimental studies are described. These two studies document linkages between land use within certain geomorphological enclaves of a drainage subcatchment and drainage water quality. Both studies were focused on nitrate nitrogen, but in general, they represent a generalised evidence of drainage runoff formation and its influence on water quality. Case study *A* was done on twenty-two tile drainage systems and their subcatchments. Case study *B* was realized on a very small 58 ha tile drained catchment Dehtáře. While case study *A* focused on monitoring the current status of land use within various drainage subcatchment zones and its relationships to drainage water quality, case study *B* was aimed at veryfying the effects of grassing in a catchment *recharge zone* on nitrate concentrations and loads in drainage waters. The main goal of both studies was to get a practical evidence for findings obtained in the CR and abroad by statistical approaches [23, 30, 44, 47] concerning the profoundly mitigative effects that grassing certain catchment areas has on the nitrate burden in drainage and surface waters.

2. Case study *A*

2.1. Materials and methods

2.1.1. Study areas

Case study *A* was carried out on 22 tile drainage systems, mostly built in slopes, which were monitored for water quality (N-NO_3) and related discharge once a fortnight for three years, 2004 – 2006. The plots studied were located in two regions of the Czech Republic (Figure 1); in the Svihov drinking water reservoir basin on the Zelivka river, and in the Southern Bohemian foothills of the Sumava mountains. The drainage systems which were selected and monitored were supposed to be functioning and not be connected to a surface water course or a pond / sewerage / wastewater outlet. All the examined drainage systems and related subcatchments were situated in crystalline complex with granite or paragneiss as parent rocks. The typical soil types were sandyloam to loamy Cambisols and loam to clayloam gleyic Stagnosols (Planosols).

For each examined drainage system, a design document (detailed construction plan) 1:1000 – 1:2000 scale was obtained and georeferrenced (rectified) to specify the exact properties and the positioning of the drainage pipes as well as to determine how the drained land and the related contributive area (subcatchment) were interconnected, using GPS and ArcGIS tools. For these areas, land use characteristics were identified either from digital cadastre maps from land registers or from LANDSAT 7 images with 30 x 30 m resolution, in both cases followed by corrections based on field surveys.

2.1.2. Determination of relative soil infiltration vulnerability

The determination of relative soil infiltration vulnerability and the delineation of infiltration-vulnerable areas was based on an analysis of five-digit codes of valuated soil ecological units (VSEU) using the method established by [38]. The VSEU maps are available as digital layers for the entire Czech Republic at a scale of 1:5 000. The VSEU code serves for the evaluation of soil characteristics according to the following criteria: main soil units, slope, exposure, skeletal character, and soil depth. Based on the categorization of these criteria, soil is classified into five relative groups according to its significance for the infiltration process, with category 1 corresponding to the maximum infiltration capacity. For the subcatchments of all tile drainage systems evaluated in this study, the proportion of land use types (arable land-AR, grassland-GR, forest-FR, built-up-BU) was determined within the infiltration–vulnerable categories I – IV; category V (the lowest infiltration–vulnerable category) was not present in any of the studied localities. An example of a tile drainage subcatchment with delineated infiltration-vulnerable areas / categories and related land use types is depicted in Figure 2.

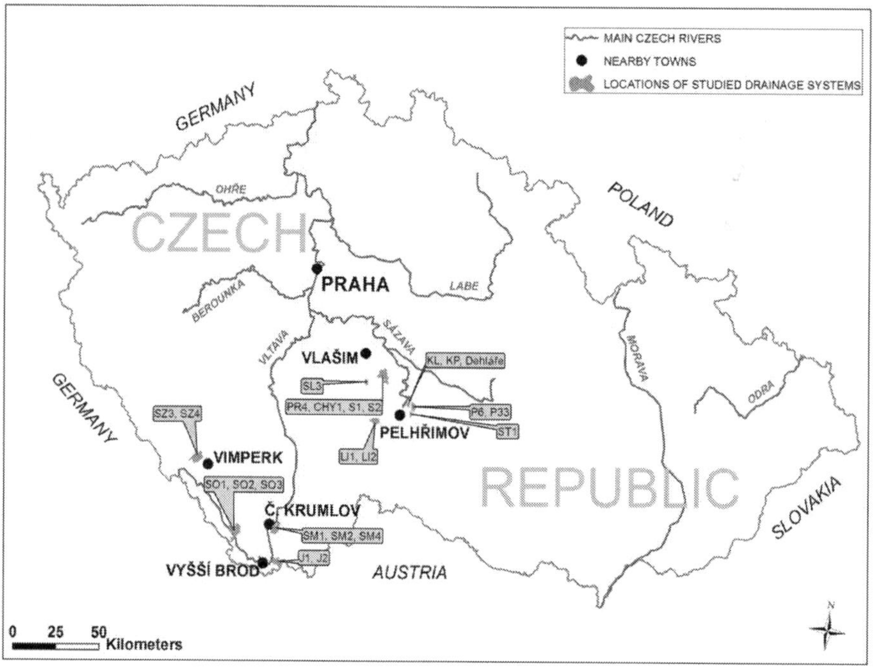

Figure 1. Location of the evaluated tile drainage systems and the Dehtáře catchment within the Czech Republic

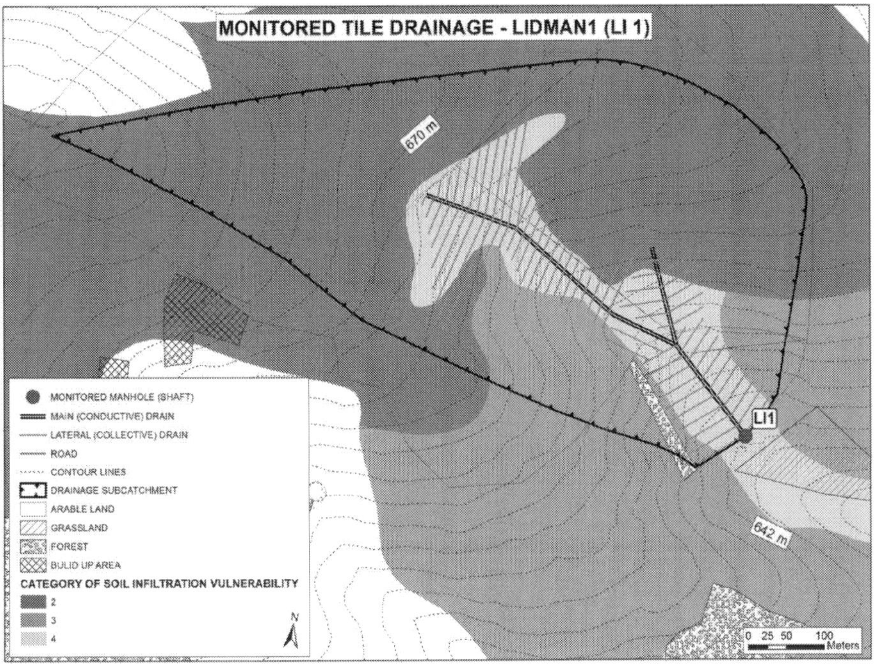

Figure 2. An example of a tile drainage subcatchment, with delineated infiltration-vulnerable areas and related land use types indicated.

2.1.3. Data preparation and employed statistical methods

Basic statistics were calculated for all monitored drainage systems. Due to the close relationships between actual nitrate concentrations and runoff [7,11] as representative values for further analyses, flow–weighted concentrations were taken in order to assess the mutual links existing solely between soil and land use catchment characteristics. Flow-weighted concentration values were computed according to the following formula:

$$Cfw = \frac{\sum (Ci * Qi)}{\sum Qi} \qquad (1)$$

Where Cfw is flow-weighted concentration

Ci is actual solute concentration in the sample

Qi is actual runoff during sample withdrawal.

To assess possible relationships between catchment soil and land use characteristics, and flow-weighetd nitrate concentration values, principal component and multiple stepwise regression (MR) methods were employed. Principal component analysis (PCA) is a widely applied method, which transforms the original data into new orthogonal coordinates in order to express the information and possible relations contained in the original data using a smaller number of variables. As the dependent variable, flow-weighted concentration (Cfw) of N-NO_3 was selected; as independent variables, the proportion of land use types within individual infiltration-vulnerable categories (AR1-4, GR1-4, FR1-4 and BU1-4). In PCA models, the following indicators were assessed: the cumulative variability explained by the first two components (%), and the component weights, showing how influential the original variables were in determining the structure of individual components. In general, factors found in the same biplot quadrants in the PCA are correlated positively; factors in opposite quadrants are correlated negatively. Factors situated around 90° angles are evaluated as independent. In the MR, the classical parameters of model correctness were considered: mutual parameter correlation – multicolinearity, autocorrelation of residuals – heteroskedasticity, detection of significant points – leverage value, deviation points – outliers (DFFITS diagnostics), and the p-values of the model and of the initial parameters. To compare different models describing identical data, the following indicators were considered: F statistics, R^2, R^2adj., (the adjusted coefficient of determination for the number of parameters, or degrees of freedom) and the Mean Square Error (MSE) [48] All analyses were done at the significance level of null hypothesis α=0.05.

2.2. Results

Table 1 shows the results of the basic statistics calculated on the tile drainage systems which were evaluated. Median N-NO_3 concentration values obtained during the period 2004-2006 were from 0.23 (SZ4) to 29.45 (KL) mg/l N-NO_3. Table 2 is an overview of land use type proportions accross four delineated soil vulnerable categories for all the evaluated drainage systems. The PCA and MR described mutual relationships between the assessed variables for the tile drainage subcatchments evaluated. In the PCA biplot (Figure 3), the points represent individual drainage subcatchments, and the beams are variables. In the assessed case, the vectors of variables Cfw and AR2 were positively correlated, as were AR3 and AR4 to a lesser extent. The factors FR4, GR4 and GR1 had a negative mutual relationship with Cfw N-$NO_{3.}$ The variables AR1 and GR3 were located in independent positions, most likely due to the very small area of all land use types in soil infiltration vulnerable category I. The results from various MR analyses are described in Table 3. The best (statistically significant and correct) model is coloured grey. It confirmed the results obtained from the PCA; the positively correlated factors were ratios of arable land within the first two soil infiltration vulnerable categories, and the ratio of grassland within the third soil infiltration vulnerable category appeared to be a mitigative factor on nitrate concentration. The final model from the MR is thus:

$$Cfw_{(N\text{-}NO3)} = 3.59 + 2.05*AR1 + 0.25*AR2 - 0.29*GR3 \quad (2)$$

Drainage system	Number of samples	Min	Max	Mean	10% Quantile	Median	90% Quantile	Flow-weighted concentration
		N-NO_3^- (mg/l)						
ST1	58	7.68	40.44	25.39	18.55	25.07	34.02	24.66
LI1	53	13.10	30.95	21.81	16.72	21.01	29.01	22.32
LI2	47	14.01	22.82	17.51	14.46	17.12	21.01	18.33
SL3	51	1.13	16.04	2.16	1.13	1.13	3.19	3.16
PR4	57	1.13	23.04	14.47	8.76	15.36	18.84	14.51
CHY1	53	18.75	39.08	27.74	23.76	27.56	33.03	30.41
P6	66	7.48	23.72	15.50	13.33	15.14	18.52	15.28
P33	66	2.71	40.89	26.89	18.51	27.56	34.00	15.58
KL	70	8.81	46.08	27.50	17.85	29.45	34.43	27.01
KP	70	9.71	43.15	23.09	13.96	23.35	30.09	25.40
S1	52	1.13	26.20	7.94	1.45	6.21	13.78	9.59
S2	52	2.26	28.46	15.42	9.98	14.46	21.46	20.67
SM1	27	1.54	24.85	19.23	16.75	20.22	22.26	19.32
SM2	27	6.30	20.20	10.81	7.40	10.55	14.29	11.45
SM4	27	5.24	22.36	17.38	14.05	17.91	20.70	15.77
SO1	27	0.35	1.89	0.82	0.47	0.79	1.12	0.72
SO2	27	0.15	6.33	2.21	1.00	1.99	3.88	1.98
SO3	25	0.65	2.23	1.71	1.13	1.88	2.14	1.43
SZ3	20	0.24	1.83	1.08	0.75	1.08	1.61	1.00
SZ4	19	0.11	1.27	0.38	0.11	0.23	0.86	0.35
J1	26	1.44	6.78	4.76	2.82	5.04	6.29	4.10
J2	26	1.54	4.09	3.01	2.12	3.03	3.87	2.78

Table 1. Basic statistics of N-NO3-concentration values in the studied drainage systems

Drainage system	Area of drainage subcatchment	Soil-vulnerable category I.				Soil-vulnerable category II.			
	ha	Arable land	Grassland	Forest	Built-Up	Arable land	Grassland	Forest	Built-Up
		AR1	GR1	FR1	BU1	AR2	GR2	FR2	BU2
		% of the whole drainage subcatchment							
ST1	34.60	0.00	0.00	0.00	0.00	78.29	0.11	1.00	0.00
LI1	28.31	0.00	0.00	0.00	0.00	79.34	0.00	0.00	0.00
LI2	24.17	0.00	0.00	0.00	0.00	76.60	0.00	0.00	0.00
SL3	8.18	0.00	0.00	0.00	0.00	47.32	36.07	0.84	0.00
PR4	9.05	0.00	0.00	0.00	0.00	71.76	0.00	0.00	0.00
CHY1	1.94	0.00	0.00	0.00	0.00	100.00	0.00	0.00	0.00
P6	15.73	0.00	0.00	0.00	0.00	68.15	0.64	0.00	0.00
P33	19.73	0.00	0.00	0.00	0.00	48.29	0.45	5.14	0.00
KL	29.44	5.37	0.28	0.00	0.00	45.29	17.32	0.00	1.32
KP	28.33	5.02	0.61	1.15	0.40	51.77	2.20	5.60	1.93
S1	0.62	0.00	0.00	0.00	0.00	0.00	0.00	0.00	0.00
S2	3.82	0.00	0.00	0.00	0.00	51.60	0.00	0.00	0.00
SM1	15.08	0.00	0.00	0.00	0.00	58.09	0.00	0.60	0.00
SM2	48.40	0.00	0.00	0.00	0.00	18.99	8.41	0.19	0.00
SM4	45.99	0.00	0.00	0.00	0.00	25.51	16.68	27.40	0.00
SO1	21.20	0.00	10.19	0.00	0.00	0.00	51.23	2.55	0.00
SO2	10.83	0.00	0.00	0.00	0.00	0.00	3.23	1.20	0.00
SO3	12.56	0.00	0.00	0.00	0.00	0.00	0.00	0.00	0.00
SZ3	52.11	0.12	6.93	0.02	0.00	14.20	9.54	13.09	0.00
SZ4	66.54	0.00	0.42	0.05	0.00	3.19	0.42	0.12	0.18
J1	46.80	0.00	5.62	1.82	0.00	0.00	54.49	10.24	0.00
J2	55.26	0.00	19.00	0.31	0.00	0.00	30.96	15.36	0.00

Drainage system	Soil-vulnerable category III.				Soil-vulnerable category IV.			
	Arable land	Grassland	Forest	Built-Up	Arable land	Grassland	Forest	Built-Up
	AR3	GR3	FR3	BU3	AR4	GR4	FR4	BU4
	% of the whole drainage subcatchment							
ST1	10.85	9.32	0.00	0.00	0.00	0.44	0.00	0.00
LI1	0.00	0.00	0.00	0.00	20.66	0.00	0.00	0.00
LI2	0.00	0.00	0.00	0.00	0.00	23.40	0.00	0.00
SL3	1.92	13.85	0.00	0.00	0.00	0.00	0.00	0.00
PR4	0.50	25.30	0.00	0.00	0.46	1.99	0.00	0.00

CHY1	0.00	0.00	0.00	0.00	0.00	0.00	0.00	0.00
P6	0.00	0.00	0.00	0.00	27.74	3.48	0.00	0.00
P33	0.00	0.00	0.00	0.00	19.92	26.21	0.00	0.00
KL	0.00	0.00	0.00	0.00	5.08	24.53	0.00	0.81
KP	1.90	0.14	0.00	0.00	24.05	5.23	0.00	0.00
S1	0.00	0.00	0.00	0.00	1.20	98.80	0.00	0.00
S2	48.40	0.00	0.00	0.00	0.00	0.00	0.00	0.00
SM1	17.71	0.00	23.61	0.00	0.00	0.00	0.00	0.00
SM2	5.95	13.35	52.93	0.00	0.00	0.19	0.00	0.00
SM4	0.00	0.00	0.00	0.00	30.42	0.00	0.00	0.00
SO1	0.00	0.00	6.89	0.00	0.00	29.15	0.00	0.00
SO2	0.00	0.00	44.23	0.00	0.00	47.92	3.42	0.00
SO3	0.00	0.00	58.20	0.00	0.00	39.17	2.63	0.00
SZ3	0.15	13.20	0.15	1.57	2.96	35.69	0.52	1.86
SZ4	8.58	0.48	66.10	0.23	0.38	17.52	2.27	0.08
J1	0.00	0.00	0.47	0.00	0.00	26.99	0.38	0.00
J2	0.00	0.00	0.00	0.00	0.00	28.23	6.03	0.11

Table 2. An overview of the evaluated tile drainage systems: Drainage subcatchment area and land use types in four soil infiltration-vulnerable categories

Variable	Coefficient	Standard deviation of residuals	P	R^2	R^2 adj.	F stat	MSE	Testing the regression triplet
Constant	3.2160		0.0378					
AR1	2.1655		0.0022					
AR2	0.2404		0.0000					
AR3	0.1157		0.1883					
GR3	-0.2684		0.0601					
Regression		4.0943	0.0000	85.9592	82.6554	26.0189	16.7630	OK
Constant	4.9037		0.0191					
AR1	1.9620		0.0053					
AR2	0.2295		0.0000					
GR3	-0.3120		0.0358					
FR4	-0.7534		0.3043					
Regression		4.1789	0.0000	85.3727	81.9309	24.8100	17.4633	autocorrel. of residuals
Constant	4.7588		0.0193					

Variable	Coefficient	Standard deviation of residuals	P	R^2	R^2 adj.	F stat	MSE	Testing the regression triplet
AR1	2.0825		0.0034					
AR2	0.2321		0.0000					
GR2	-0.0610		0.3254					
GR3	-0.2802		0.0546					
Regression		4.1900	0.0000	85.2946	81.8345	24.6509	17.5565	OK
Constant	3.5890		0.0223					
AR1	2.0512		0.0035					
AR2	0.2471		0.0000					
GR3	-0.2892		0.0469					
Regression		4.1930	0.0000	84.4073	81.8086	32.4800	17.5815	OK
Constant	2.9001		0.0723					
AR1	2.2738		0.0026					
AR2	0.2363		0.0000					
Regression		4.5681	0.0000	80.4646	78.4082	39.1296	20.8679	OK

Table 3. A list of statistically significant and correct multiple regression models (the grey band indicates the best regression model).

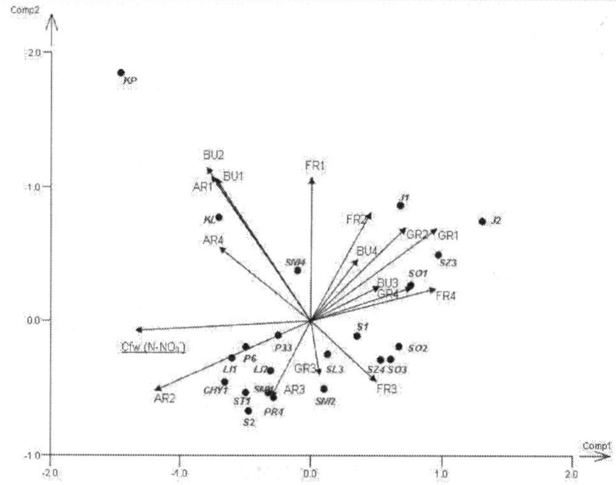

Figure 3. PCA Biplot for all variables – land use types in soil-infiltration vulnerable categories (I-IV)

Based on the acquired results, it can be said that the most important influence on values of nitrate nitrogen in drainage water was, in the case of the drainage systems analysed, the ratio (%) of ploughed land within the most infiltration vulnerable catchment areas.

3. Case study *B*

3.1. Materials and methods

3.1.1. Study area

The experimental catchment Dehtáře (Figure 1 and Figure 4) is situated in the Bohemian-Moravian Highlands, Czech Republic. It is a locally typical small agricultural catchment, where the tile drainage acts as the only permanent runoff from the catchment and the drainage system was built in the slope. The area is 57.9 ha, with tile-drained areas occupying 19 ha (32%). It has been used mainly as agricultural land, with low forest representation. The agricultural land is mostly exploited as arable, with permanent grassland in the lower part of the catchment. The altitude varies between 549.8 and 497 m a.sl. Total precipitation throughout the vegetation period ranges between 350 and 450 mm, and in the winter months between 250 and 300 mm, with a total annual average of 666 mm. The substrate is formed by partially migmatized paragneiss in various degrees of degradation. Quaternary sediments are represented by slope sands and loams reaching 1-2 m thickness. The representation of soils is variable, with Gleyed Cambisols, Gleysols, and sporadically Histosols. In the recharge area, the soil cover is more homogenous, with Haplic and, Shallow Haplic Cambisols and Cambic Hyperskeletic Leptosol prevailing. The drainage system was built in 1977, with a slope of 5%. The spacing of collection drains is 13 or 20 m apart; the depth of collection drains is 1.0 m, of conduit drains 1.1 m, and the interception drains are deposited at a 1.1-1.8 m depth. Detailed information about the catchment including a geophysical survey was published e.g. by [49].

3.1.2. Design of the pilot plant experiment

The water quality has been monitored since 2003. Samples have been taken at one or two week intervals. Five sites in the drainage system, with different land use in recharge and discharge areas, were chosen to be monitored for this analysis (Table 4). Part of the recharge area (Figure 4), with an area of 4.6 ha, has been grassed since the hydrologic year 2007. To evaluate the effects of this grassing, the whole time monitored was divided into two periods, period 1-before grassing (2003-2006) and period 2-after grassing (2007-2013). All grassland in the catchment was fertilised by approximately 100 kg N/ha per season (mostly by urea and liquid manure). The arable land in the catchment was fertilised according to crop rotation (grains, potatos and oilseed rape) in the amount of cca. 120 kg N/ha per season.

The most important sites for the evaluation were K1, K2 and K5. The site K1 represents an outlet of an eponymous systematic drainage subsystem with 1 ha area and drains spaced at 13 m. This subsystem is connected to an intercepting drain K2 which collects hypodermic water from upslope areas. The discharge area of these sites (area of their placement) was permanently

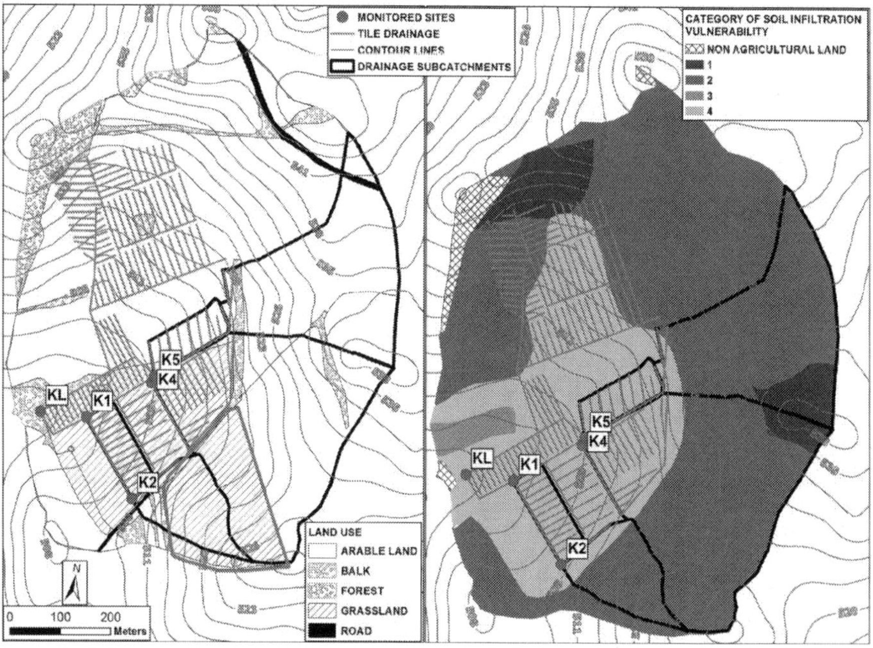

Figure 4. An overview map of the experimental catchment Dehtáře. Red encircled is the area experimentally grassed in 2007.

grassed, their recharge area had been used as arable land until the hydrologic year 2007, when it was experimentally grassed. The site K5 represents an outlet of an eponymous systematic drainage subsystem with 1 ha area and drains spaced at 20 m; both recharge and discharge areas were used as arable land. The site KL is the closing outlet of all these subsystems.

Site	Drainage type	Period 1 (2003-2006) land use		Period 2 (2007-2013) land use	
		Discharge area	Recharge area	Discharge area	Recharge area
K1	systematic	grassland	arable land	grassland	grassland
K2	intercepting	intercepting drain	arable land	intercepting drain	grassland
K4	systematic	grassland	arable land	grassland	arable land
K5	systematic	arable land	arable land	arable land	arable land
KL	closing profile	arable land + grassland	arable land	arable land + grassland	arable land + grassland

Table 4. Land use in the recharge and discharge areas of the drainage systems used for the experiment

http://dx.doi.org/10.5772/59298

3.1.3. *Data evaluation*

Similarly to case study *A*, the solute concentration value data were adjusted to facilitate hydrologic modeling. Flow-weighed nitrate concentration values, computed according to formula (1), were further analysed using the statistical software Statgraphics. In addition to calculating common summary statistics, the analysis of variance and Kruskall-Wallis test were conducted to test whether there were significant changes in nitrate concentrations and nitrogen load caused by grassing.

3.2. Results

The nitrate concentrations in drainage waters were strongly variable during the entire period monitored and also between the particular periods. The observed concentrations varied from 18 to 253 mg/l throughout the whole period monitored. The main reasons for this variability were the variable soil nitrogen stocks and the strong concentration dependence on the discharge levels. That is why the highest concentrations were measured in late summer or early autumn – during the period of low drainage discharges and prevailing base flow. On the contrary, during spring snowmelt and summer high-flow events with increased discharge (especially its direct-event component), nitrate concentrations decreased due to a high degree of dilution. The exceptions were some high-flow events measured on sites with arable land in the recharge zone just after the application of fertilizer. The inter-seasonal variability was caused mainly by different precipitation courses in particular seasons and by crop rotation.

The results of the statistical analysis for both periods of the experiment (2004-2006 and 2007-2011) are depicted in Figure 5 and Table 5. The evaluation showed that the flow-weighted nitrate concentrations in period 1 (before grassing the recharge zone) were surprisingly higher in drainage subsystems K1 and K4 with the permanent grassland in drained area (discharge zone) than in the subsystem under arable land. Moreover, the concentrations mostly exceeded the level of 100 mg/l. After grassing the K1 subsystem recharge area, some changes occurred. At first, the nitrate concentration decreased during high–flow events. After that, approximately one year after grassing, the long-term course of NO_3 concentrations changed direction and became decreasing (Figure 6).

The significance of the changes in nitrate concentration values was tested using the Kruskall-Wallis test comparing the medians of the concentrations from periods 1 and 2 (before and after grassing). The results of this test are presented in Table 6. It is obvious that the statistically significant decrease in nitrate concentrations happened in the grassed recharge zone. Decreases of 32.1% and 25.7% were detected in systematic drainage subsystem K1 and intercepting drain K2 respectively. In the same period, an increase in nitrate concentration was detected in sites without land use change in their recharge zone. There was an increase of 10.8% in the drainage subsystem K5 with arable land in both (recharge and discharge) zones and of 8.6% in the subsystem K4 with grassland in the discharge zone, but arable land in the recharge zone. Evaluating the whole drainage system, the fall in nitrate concentrations by 10.5% was detected after grassing about 20% of this systems recharge zone. Nevertheless, from the statistical point of view, this fall appears to be insignificant.

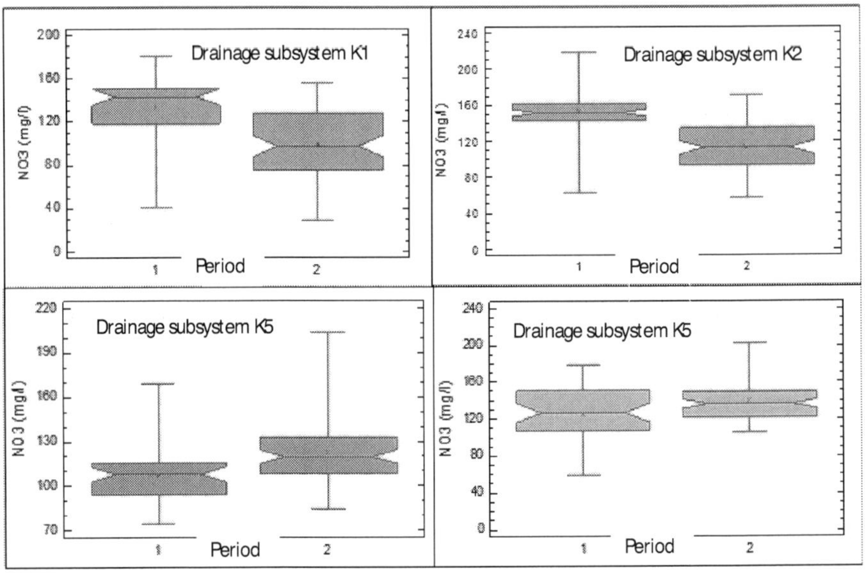

Figure 5. Nitrate concentrations in waters of drainage subsystems with different landuse before (period 1) and after (period 2) grassing of a part of a catchment recharge zone.

Site	Period 1 NO_3 (mg/l)				Period 2 NO_3 (mg/l)			
	Mean	Median	Minimum	Maximum	Mean	Median	Minimum	Maximum
K1	132.7	143.3	40.6	181.5	99.5	97.4	29.2	155.9
K2	152.5	151.5	62.9	219.0	113.0	112.9	56.7	170.5
K5	107.9	108.8	74.0	168.6	122.7	119.7	84.7	203.8
K4	125.3	126.6	58.6	178.2	139.4	136.2	105.6	202.9
KL	109.4	107.3	39.7	168.5	98.6	96.0	36.1	147.9

Table 5. The basic statistical evaluation of nitrate concentrations in water of drainage subsystems with different landuse before (period 1) and after (period 2) grassing in part of a recharge zone.

Despite the decrease proved above, the nitrate concentration values in the drainage water have still remained too high. The reason is that most of the samples were taken during prevailing base flow. The concentrations in this runoff component would remain high for a longer time because of the long residence time of ground and hypodermic water in the catchment. However, what is much more important than the instantaneous level of nitrate concentration, is that the trend in nitrate concentration became permanently decreasing in all measured sites

http://dx.doi.org/10.5772/59298

with grassed recharge zone (Figure 6). While the linear trend was detected increasing in all sites during period 1, it reversed approximately one year after grassing.

Site	Landuse in recharge zone during experiment	Change in median of NO_3 concentrations (%)	Results of Kruskall-Wallis test	p value
K1	grassland	-32.1	significant	0.001
K2	grassland	-25.7	significant	0.001
K5	arable land	10.8	significant	0.002
K4	arable land	8.6	significant	0.028
KL	arable land + grassland	-10.5	insignificant	0.065

Table 6. The rate of change in nitrate concentrations after partially grassing a recharge zone and the results of the Kruskall-Wallis test for determining the significance of the change

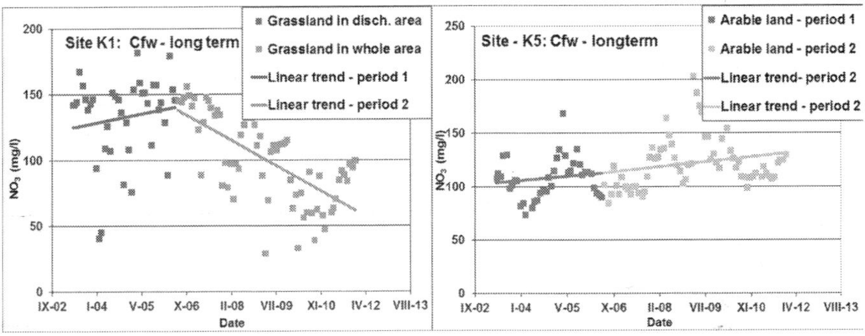

Figure 6. The long term trend of nitrate concentrations in the drainage subsystem K2 with grassed recharge zone and drainage subsystem K5 with remaining arable land in its recharge zone

In association with the change in nitrate concentrations, the nitrate-nitrogen leaching decreased after grassing in recharge area of the drainage system. Basic statistical evaluation of nitrogen leaching from drained subcatchments with different land use is depicted in Figure 7. Again, in all sites with grassed recharge zone, the decrease in all statistical characteristics of nitrogen load happened since part of the recharge zone was grassed (period 2). In the scale of whole drainage system, the monthly average load decreased by 23% from 3.2 kg N/month/ha to 2.6 kg N/month/ha. In the drainage subsystem K1, where the recharge zone was grassed completely, the decrease of the monthly average nitrogen load was even by 47% from 4.75 kg N/month/ha in period 1 to 2.52 kg N/month/ha in period 2. Evaluating drainage subsystems without land use change in the recharge zone, N load stagnation was registered in subsystem K4 (grassland in the discharge zone and arable land in the recharge zone), where the average monthly N load was 4.0 kg N/month/ha in period 1 and 3.9 N/month/ha in period 2. In the subsystem K5 (arable land in both zones), the increase of N load by 17% (from 4.1 to 4.8 kg N/month/ha) was recorded consequently with N load decrease in subsystems with grassed recharge zones. This decrease manifested as much more significant during high–flow events, which had the biggest share of the total annual nitrogen loads.

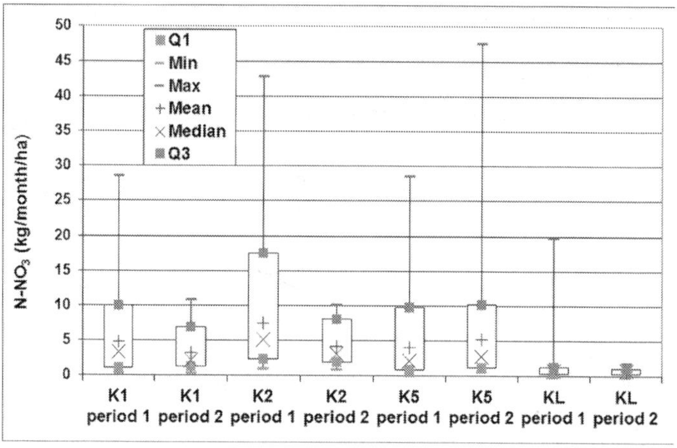

Figure 7. The average monthly N-NO_3 load per 1 ha from drainage subsystems with different landuse (period 1 – before grassing, period 2-after partially grassing the recharge zone)

4. Discussion

The results of both studies reported in this chapter corresponds with findings about grassland ability to reduce nitrogen leaching, as reported by many author from the Czech Republic and other countries. As it has been often demonstrated [30,31,45,50,51], nitrate concentrations in waters generally increase in connection with land use gradient from forests, meadows, pastures to arable lands; the last being the most worsening one. According to [44], the nitrate concentrations in small water courses of the Czech Republic are more affected by the area of arable land within their catchments than by the amount – to a certain level-of applied nitrogenous fertilizers. Other authors [52,53,54] have also reported the positive effect of grassing on water quality while some authors mention a certain lag effect of grassing on water quality improvement [55]. The degree of NO_3 concentration and N leaching changes evaluated by the case study *B* is within the range of findings reported by [44] in the Czech Republic. That work documented by multifactorial regression within the Švihov drinking water reservoir basin the share of arable lands as the most important factor affecting the nitrate concentration in water of small water courses. Similarly, the study [23] proved for catchments of three different scales significant linkages between landuse and water quality, reporting that every decrease in arable land area by 10% would cause an average decrease in C90 NO_3 values of 6.4 mg/l. The work [44] further proved that for the small streams in the Šumava region (Southern CR), improvement of water quality in this region was caused by grassing of arable and tile drained soils. The delay of reaction of nitrate concentration on grassing, detected in the case study *B*, was approximately one year. This period corresponds to the average residence time of the prevailing drainage runoff component (hypodermic water) in drainage discharge in sloped catchments, as reported e.g. by [56].

The ability of grasslands to reduce nitrate pollution is explained by the fact that grassland can absorb and use bigger amount of nitrogen in comparison to field crops [57]. Permanent grasslands cover the soil year round and have a big stock of active subsurface biomass in the root system, which can immobilize a significant amount of soil nitrogen. Nitrogen in mineral form thus exists in the soil in small concentration because nitrates produced by nitrification are readilly utilized by the grass. Besides this, after grassland is fertilised, nitrogen is quickly immobilised in the soil organic matter and protected againts leaching. Another important factor for grassland efficiency in nitrogen load is the bigger amount and increased activity of soil microbes, which is much higher under grassland than under field crops [58-61].

The results of the experiment *B* in the Dehtáře catchment also implied that grassing, being considered as an efficient measure for nitrogen leaching mitigation, should include a proper targeting in certain landscape zones, especially in tile drained catchments [8,44,62]. Drainage systems in the foothill areas of highland are characterized by their location in slopes and a considerable portion of drainage runoff can originate outside of the drained area [18-19,63]. To understand the mechanism of drainage runoff generation in sloped conditions, the theory of catchment slope zones (areas) must be applied. Perceived hydrogeologically, a catchment splits into recharge zones, where rainwater infiltrates and gradually joins groundwater, and discharge zones, where groundwater approaches surface water body or soil surface [22]. The recharge zones are usually situated in the morphologically uppermost areas of a catchment, close to the catchment watershed divide. The soils of recharge zones are typically shallow and stony, with high sand content and high infiltration capacity. The coarse-textured soils of the recharge zones are, with a respect to groundwater resources, well suited to grassing, which, beside water quality benefits, increases their field water holding capacity and enables infiltration of a bigger precipitation amounts – compared to ploughlands, including rainstorms [18,23,64-65]. The discharge zones are usually situated in the foothills and along surface water courses and lakes and are prone to surface waterlogging. Typical soils in the discharge zones are generally deeper and heavier, with higher clay content and a lower capacity for infiltration. A connection between the recharge zones and the discharge zones is carried out by transient zones, where water from precipitation is either transformed to surface runoff or to groundwater which flows downslope in a quasi-steady way [18,24,27]. The transient zones are situated mainly in the middle sections of slopes. Groundwater in natural catchments flows from the recharge zones to the discharge zones. The discussed drainage systems in slopes are mostly placed in at the interface of the transient and discharge zones or in the discharge zones [18]. The joint influence of different land use, soil properties and tile drainage within various landscape zones on the hydrological regime of a small catchment was demonstrated by [2]. Being built in slope, tile drainage systems represent a shortcut between recharge and discharge zones which significantly curtails the water residence time in catchments, hastens the runoff reaction to precipitation, shortens the time to reach the peak discharge during events and increase the nitrate concentrations and loads in related waters [4,11,17,66-68].

5. Conclusions

The results presented show that nitrate concentration values in drainage water were influenced the most by the land use of the recharge zones within the drainage subcatchment. These findings can be generalised for slopy agricultural catchments with common land use in soil environments formed on crystalline rocks. The land use and soil analyses together with monitoring the drainage water quality and quantity, and also an experiment with land use change in the recharge zone proves that grassing focused on the proper catchment area can be employed as a useful tool for reducing nitrates in drainage water. While permanent grassland placed directly in the drained area (corresponding to the catchment discharge zone) did not show any influence, the grassing focused on the catchment recharge area demonstrated a significant decrease in both, NO_3 concentrations and N loads. The acquired findings are of crucial importance for improving the water quality of small streams as well as groundwater in agriculturally exploited areas, for planning protective zones within large catchments of potable water reservoirs, and also for protecting small local surface or groundwater sources of potable water.

Acknowledgements

The results published in this chapter were obtained with financial support from the Research Programme No. 0002704902 and from the Czech National Agency for Agricultural Research – Project No. QI 111C034. The authors wish to thank Rebecca Hollinger for language corrections and Jana Maxová for technical help.

Author details

Petr Fučík[1], Antonín Zajíček[1], Renata Duffková[1] and Tomáš Kvítek[2]

1 Research Institute for Soil and Water Conservation, Department of Hydrology and Water Protection, Prague, Czech Republic

2 University of South Bohemia in České Budějovice, Faculty of Agriculture, Czech Republic

References

[1] Blann KL, Anderson JL, Sands GR, Vondracek B. Effects of agricultural drainage on aquatic ecosystems: A review. Crit. Rev. Environ. Sci. Technol 2009;39(11): 909–1001.

[2] Duffková R, Zajíček A, Nováková E. Actual evapotranspiration from partially tile drained fields as influenced by soil properties, terrain and crop. Soil and Water Res. 2011;6: 131-146.

[3] Rahman MM, Lin Z, Jia X, Steele DD, DeSutter T. Impact of subsurface drainage on streamflows in the Red River of the North basin. Journal of Hydrology 2014;511: 474–483. http://dx.doi.org/10.1016/j.jhydrol.2014.01.070.

[4] Robinson M. Impact of Improved Land Drainage on River Flows. Report No. 113. Wallingford: Institute of Hydrology; 1990.

[5] Ahiablame LM, Chaubey I, Smith DR, Engel BA. Effect of tile effluent on nutrient concentration and retention efficiency in agricultural drainage ditches. Agricultural Water Management 2011;98: 1271–1279. doi:10.1016/j.agwat.2011.03.002.

[6] Brown CD, van Beinum W. Pesticide transport via sub-surface drains in Europe. Environmental Pollution 2009;157: 3314–3324. doi:10.1016/j.envpol.2009.06.029.

[7] Fučík P, Kaplická M, Kvítek T, Peterková J. Dynamics of stream water quality during snowmelt and rainfall–runoff events in a small agricultural catchment. Clean Soil Air Water 2012;40: 154–163. doi:10.1002/clen.201100248.

[8] Zajíček A, Kvítek T, Kaplická M, Doležal F, Kulhavý Z, Bystřický V, Žlábek P. Drainage water temperature as a basis for verifying drainage runoff composition on slopes. Hydrological processes 2011;25: 3204-3215. doi: 10.1002/hyp.8039.

[9] Kröger R, Pierce SC, Littlejohn KA, Moore MT, Farris JL. Decreasing nitrate-N loads to coastal ecosystems with innovative drainage management strategies in agricultural landscapes: An experimental approach. Agricultural Water Management 2012;103: 162–166. doi:10.1016/j.agwat.2011.11.009.

[10] Morrison J, Madramootoo CA, Chikhaoui M. Modeling the influence of tile drainage flow and tile spacing on phosphorus losses from two agricultural fields in southern Québec. Water Quality Research Journal of Canada 2013;48(3) 279–293. doi:10.2166/wqrjc.2013.053

[11] Tiemeyer B, Frings J, Kahle P, Köhne S, Lennartz B. A comprehensive study of nutrient losses, soil properties and groundwater concentrations in a degraded peatland used as an intensive meadow – Implications for re-wetting. J.Hydrol. 2007;345: 80–101.

[12] Fučík P, Novák P, Žížala D. A combined statistical approach for evaluation of the effects of land use, agricultural and urban activities on stream water chemistry in small tile-drained catchments of south Bohemia, Czech Republic. Environmental Earth Sciences 2014. doi: 10.1007/s12665-014-3131-y.

[13] Heilman P et al. Extending results from agricultural fields with intensively monitored data to surrounding areas for water quality management. Agricultural Systems 2012;106: 59–71. doi:10.1016/j.agsy.2011.10.010.

[14] Hirt U, Hammann T, Meyer BC. Mesoscalic estimation of nitrogen discharge via drainage systems. Limnologica 2005;35: 206–219. doi:10.1016/j.limno.2005.06.005.

[15] Kennedy CD et al. Dynamics of nitrate and chloride during storm events in agricultural catchments with different subsurface drainage intensity (Indiana, USA). Journal of Hydrology 2012;466–467: 1–10. http://dx.doi.org/10.1016/j.jhydrol.2012.05.002.

[16] Kulhavý Z et al. Management of agricultural drainage systems in the Czech Republic. Irrigation and Drainage 2007;56: 141-149.

[17] Turunen M, Warsta L. et al. Modeling water balance and effects of different subsurface drainage methods on water outflow components in a clayey agricultural field in boreal conditions. Agricultural Water Management 2013;121: 135-148.

[18] Doležal F, Kvítek T. The role of recharge zones, discharge zones, springs and tile drainage systems in peneplains of Central European highlands with regard to water quality generation processes. Physics and Chemistry of the Earth 2004;29: 775-785.

[19] Herrmann A, Duncker D. Runoff formation in a tile-drained agricultural basin of the Harz Mountain Foreland, Northern Germany. Soil and Water Research 2008;3(3) 83–97.

[20] Fučík P, Hejduk T, Peterková J. Quantifying Water Pollution Sources in a Small Tile-drained Agricultural Watershed. Clean Soil Air Water 2014. doi: 10.1002/clen.201300929.

[21] Honisch M, Hellmeier C, Weiss K. Response of surface and subsurface water quality to land use changes. Geoderma 2002;105(3) 277–298. doi: 10.1016/S0016-7061(01)00108-2.

[22] Serrano ES. Hydrology for Engineers, Geologists and Enviromental professionals. Lexington, Kentucky: HydroScience Inc.; 1997. ISBN 0-9655643-9-8.

[23] Fučík P, Kvítek T, Lexa M, Novák P, Bílková A. Assessing the Stream Water Quality Dynamics in Connection with Land Use in Agricultural Catchments of Different Scales. Soil & Water Res 2008;3: 98–112.

[24] Zheng FL, Huang ChH, Norton LD. Effects of Near-Surface Hydraulic Gradients on Nitrate and Phosphorus Losses in Surface Runoff. Journal of Environmental Quality 2004;33: 2174–2182.

[25] Barrett ME, Charbeneau RJ. A parsimonious model for simulating flow in a karst aquifer. Journal of Hydrology 1997,196: 47–65.

[26] Minár J, Evans S. Elementary forms for land surface segmentation: The theoretical basis of terrain analysis and geomorphological mapping. Geomorprology 2008;95: 236–259.

[27] Reprinted with permission from [Duffková R. 2013. Influence of Soil Physical Properties and Terrain Relief on Actual Evapotranspiration in the Catchment with Prevail-

ing Arable Land Determined by Energy Balance and Bowen Ratio, in Evapotranspiration - An Overview, Stavros G. Alexandris, Ruzica Sticevic (Eds.), pp 207 – 226. ISBN: 978-953-51-1115-3, InTech, DOI: 10.5772/52810.]

[28] Dixon B. Prediction of ground water vulnerability using an integrated GIS-based neuro-fuzzy techniques. Journal of Spatial Hydrology 2004;4(2): 1-38.

[29] Dragon K. Groundwater nitrate pollution in the recharge zone of a regional Quaternary flow system (Wielkopolska region, Poland). Environ Earth Sci 2013;68: 2099–2109.

[30] Edwards AC, Pugh K, Wright GG, Sinclair AH, Reaves GA.Nitrate status of two major rivers in N. E. Scotland with respect to land use and fertiliser additions. Chemistry and Ecology 1990;4: 97-101.

[31] Lord EI, Johnson PA, Archer JR. Nitrate Sensitive Areas: a study of large scale control of nitrate loss in England. Soil Use and Management 1999;15: 201-207.

[32] Raposo JR, Molinero J, Dafonte J. Parameterization and quantification of recharge in crystalline fractured bedrocks in Galicia-Costa (NW Spain). Hydrol. Earth Syst. Sci. 2012;16: 1667–1683.

[33] Ross M et al. Assessing rock aquifer vulnerability using downward advective times from a 3D model of surficial geology: A case study from the St. Lawrence Lowlands, Canada. Geofísica Internacional 2004; 43(4) 591-602.

[34] Aller L, Bennet T, Lehr JH, Petty RJ, Hackett G. DRASTIC: A standardised system for evaluating groundwater pollution potential using hydrogeological settings. EPA/600/2–87/035. Oklahoma: US Environmental Protection Agency, Agency Ada; 1987.

[35] Civita MV. The Combined Approach When Assessing and Mapping Groundwater Vulnerability to Contamination. J. Water Resource and Protection 2010;2: 14–28.

[36] Goldscheider N, Klute M, Sturm S, Hotzl H. The PI Method: a GIS Based Approach to Mapping Groundwater Vulnerability with Special Consideration of Karst Aquifers, Z. Angew. Geol. 2000;3: 157–166.

[37] Vias JM, Andreo B, Perles MJ, Carrasco F, Vadillo I, Jimenez P. Proposed Method for Groundwater Vulnerability Mapping in Carbonate (Karstic) aquifers: the COP method: Application in Two Pilot Sites in Southern Spain. Hydr. J. 2006;6: 1–14.

[38] Janglová R, Kvítek T Novák P. Soil infiltration capacity categorisation based on a geo-informatic synthesis of the Comprehensive Soil Survey and Valuated Soil-Ecological Units data. Soil and Water 2003;2: 61-82. Prague; Research Institute for Soil and Water Conservation: 2003.

[39] Kvítek T, Fučík P, Novák P, Novotný I, Kaplická M, Žížala D. Identifikace potenciálních zdrojových lokalit plošného zemědělského znečištění-standardizovaný podklad pro projektování komplexních pozemkových úprav. (Identification of Potential Source Areas of Non-Point Agricultural Pollution – A Standardized Methodics for

Land Adjustment and Consolidation). Prague; Research Institute for Soil and Water Conservation: 2008. 34 p., ISBN 978-80-904027-3-7.

[40] Novák P et al. Metodický postup tvorby syntetické mapy zranitelnosti podzemních vod. Uplatněná certifikovaná metodika. (Synthetic Map of Groundwater Vulnerability Assessment: A certified methodics). Prague; Research Institute for Soil and Water Conservation: 2012. 44 p. ISBN 978-80-87361-19-1.

[41] Tomer MD, Moorman TB, Rossi CG. Assesment of the Iowa River´s South fork watershed: part 1. Water quality. Journal of Soil and Water Conservation 2008;63(6) 360-370.

[42] Haberle J, Káš M. Simulation of nitrogen leaching and nitrate concentration in a long-term field experiment. Journal of Central European Agriculture 2012;3: 416-425.

[43] Schilling K, Spooner J. Effects of Watershed-Scale Land Use Change on Stream Nitrate Concentrations. J. Environ. Qual. 2006;35: 2132–2145. doi:10.2134/jeq2006.0157.

[44] Kvítek T, Žlábek P, Bystřický V, Fučík P, Lexa M, Gergel J, Novák P, Ondr P. Changes of nitrate concentrations in surface waters influenced by land use in the crystalline complex of the Czech Republic. Physics and Chemistry of the Earth 2009;34: 541–551.

[45] Reynolds B, Edwards AC. Factors influencing dissolved nitrogen concentrations and loading in upland stream of the UK. Agricultural water management 1995;27: 181-202.

[46] Whitmore AP, Bradbury NJ, Johnson PA. Potential contribution of ploughed grassland to nitrate leaching. Agriculture, Ecosystems and Environment 1992;39: 221-233.

[47] Thornton GJP, Dise NB. The influence of catchment characteristics, agricultural activities and atmospheric deposition on the chemistry of small streams in the English Lake District. The Science of the Total Environment 1998;216: 63–75.

[48] Helsel DR, Hirsch RM. Statistical methods in water resources. U.S. Geological Survey Techniques of Water Resources Investigations, book 4, chap. A3, p 524: 2002. http://pubs.usgs.gov/twri/twri4a3/

[49] Nováková E, Karous M, Zajíček A, Karousová M. Evaluation of ground penetrating radar and vertical electrical sounding methods to determine soil horizons and bedrock at the locality Dehtáře. Soil & Water Res. 2013;8(3) 105–112.

[50] Ruiz L, Abiven S, Durand P, Vertès F, Beaujouan V. Effect on nitrate concentration in stream water of agricultural practices in small catchments in Brittany: I. Annual nitrogen budgets. Hydrology and Earth System Sciences 2002;6(3) 497-505.

[51] Worrall F, Burt T, Adamson J. Controls on the chemistry of runoff from an upland peat catchment. Hydrol. Process. 2003;17: 2063–2083.

[52] Strock JS, Porter PM, Russelle MP. Cover cropping to reduce nitrate loss through subsurface drainage in the northern U.S. Corn Belt. Journal of Environmental Quality 2004;33(3) 1010-1016.

[53] Kaspar TC, Jaynes DB, Parkin TB, Moorman TB, Singer JW. Effectiveness of oat and rye cover crops in reducing nitrate losses in drainage water. Agricultural Water Management 2012;110: 25– 33.

[54] Heggenstaller AH, Anex RP, Liebman M, Sundberg DN, Gibson LR. Productivity and nutrient dynamics in bioenergy double-cropping systems. Agronomy journal 2008;100: 1740–1748. doi:10.2134/agronj2008.0087.

[55] Meals DW, Dressing SA, Davenport TE. Lag time in water quality response to best management practices: A review. J Environ Qual 2010;39(1) 85–96. doi:10.2134/jeq2009.0108.

[56] Bůzek F, Bystřický V, Kadlecová R, Kvítek T, Ondr P, Šanda M, Zajíček A, Žlábek P. Application of two-component model of drainage discharge to nitrate contamination. Journal of Contaminant Hydrology 2009;106: 99–117.

[57] Whitehead DC. Grassland nitrogen. Wallingford: CABI Publ.; 1995.

[58] Marschner B, Kalbitz K. Controls of bioavailability and biodegradability of dissolved organic matter in soils. Geoderma 2003;113: 211-235.

[59] Carpenter SR, Caraco NF, Corell DL, Howarth RW, Sharpley AN, Smith VH. Non point polution of surface waters with phosphorus and nitrogen. Ecological Applications 1998;8: 559-568.

[60] Merino A., Pérez-Batallón P., Macías F. Responses of soil organic matter and greenhouse gas fluxes to soil management and land use changes in a humid temperate region of southern Europe. Soil Biology & Biochemistry 2004;36: 917–925.

[61] Griffiths BS, Hallett PD, Kuan HL, Gregory AS, Watts CW, Whitmore AP. Functional resilience of soil microbial communities depends on both soil structure and microbial community composition. Biol. Fertil. Soils 2008;44: 745–754.

[62] Lemke AM, Kirkham KG, Lindenbaum TT, Herbert ME, Tear TH, Perry WL, Herkert JR. Evaluating Agricultural Best Management Practices in Tile-Drained Subwatersheds of the Mackinaw River, Illinois. J Environ Qual. 2011;40(4) 1215-28. doi: 10.2134/jeq2010.0119.

[63] Stone M, Krishnappan BG. In: Serrano ES. Hydrology for Engineers, Geologists and Enviromental professionals. Lexington, Kentucky: HydroScience Inc.; 1997.

[64] Constantin J, Beaudoin N, Launay M, Duval J, Mary B. Long-term nitrogen dynamics in various catch crop scenarios: Test and simulations with STICS model in a temperate climate. Agriculture, Ecosystems and Environment 2012;147: 36– 46.

[65] Laurent F, Ruelland D. Assessing impacts of alternative land use and agricultural practices on nitrate pollution at the catchment scale. Journal of Hydrology 2011;409: 440–450.

[66] Billy C, Birgand F, Ansart P, Peschard J, Sebilo M, Tournebize J. Factors controlling nitrate concentrations in surface waters of an artificially drained agricultural watershed. Landscape Ecol. 2013;28:665–684.

[67] Tomer MD, Wilson CG, Moorman TB, Cole KJ, Heer D, Isenhart TM. Source-Pathway Separation of Multiple Contaminants during a Rainfall-Runoff Event in an Artificially Drained Agricultural Watershed. J. Environ. Qual. 2010;39: 882–895.

[68] Tachecí P, Žlábek P, Kvítek T, Peterková J. Analysis of Rainfall-Runoff Events in Four Subcatchments of the Kopaninský Potok (Czech Republic). Bodenkultur 2013;64 (3–4) 105–111.